Addressing Climate Anxiety in Schools

This monograph presents a contemporary examination of climate anxiety within schools. Featuring contributions from experts across Canada, Austria, Australia, New Zealand, the United Kingdom, Italy, and Finland, the book underscores the prevalence of climate anxiety, a phenomenon often overlooked in discussions about climate change and education. The monograph is divided into two sections. The first section begins by outlining how climate anxiety manifests in schools, examining the theoretical underpinnings of climate change education and its psychological impact on students and teachers. The second section presents innovative and practical strategies for mitigating climate anxiety in the classroom, highlighting the importance of cohesive learning environments and cross-curricular approaches. Readers will benefit from the book's international perspective and its blend of theory and practice, gaining valuable insights into how to address climate anxiety and foster resilience in educational contexts.

An international, empirical, and ethnographic evidence-based perspective of climate anxiety in classroom, this book will appeal to scholars, researchers, postgraduate students, and educators with interests in climate change education, sustainability education, policy and administration, mental health, and pedagogy.

Julie K. Corkett is Full Professor at the Schulich School of Education, Nipissing University, Canada.

Wafaa Mohammed Moawad Abd-El-Aal is Assistant Professor in the College of Education at Sultan Quaboos University, Sultanate of Oman, and Associate Professor in the Faculty of Education at Beni-Suef University, Egypt.

Astrid Steele is retired Associate Professor of Science and Environmental Education from the Schulich School of Education, Nipissing University, Canada.

Research and Teaching in Environmental Studies

This series brings together international educators and researchers working from a variety of perspectives to explore and present best practice for research and teaching in environmental studies.

Given the urgency of environmental problems, our approach to the research and teaching of environmental studies is crucial. Reflecting on examples of success and failure within the field, this collection showcases authors from a diverse range of environmental disciplines including climate change, environmental communication and sustainable development. Lessons learned from interdisciplinary and transdisciplinary research are presented, as well as teaching and classroom methodology for specific countries and disciplines.

Transformative Sustainability Education
Reimagining Our Future
Elizabeth A. Lange

Poetry and the Global Climate Crisis
Creative Educational Approaches to Complex Challenges
Edited by Amataritsero Ede, Sandra Lee Kleppe, and Angela Sorby

Research Journeys to Net Zero
Current and Future Leaders
Edited by Kyungeun Sung, Patrick Isherwood, and Richie Moalosi

Polar and Climate Change Education
Citizen Science and Sustainability
Edited by Gisele M. Arruda

Addressing Climate Anxiety in Schools
Pedagogical Perspectives and Theoretical Foundations
Edited by Julie K. Corkett, Wafaa Mohammed Moawad Abd-El-Aal, and Astrid Steele

For more information about this series, please visit: www.routledge.com/Research-and-Teaching-in-Environmental-Studies/book-series/RTES

Addressing Climate Anxiety in Schools

Pedagogical Perspectives and Theoretical Foundations

Edited by Julie K. Corkett, Wafaa Mohammed Moawad Abd-El-Aal, and Astrid Steele

Routledge
Taylor & Francis Group

NEW YORK AND LONDON

First published 2025
by Routledge
605 Third Avenue, New York, NY 10158

and by Routledge
4 Park Square, Milton Park, Abingdon, Oxon, OX14 4RN

Routledge is an imprint of the Taylor & Francis Group, an informa business

ISBN: 978-1-032-75752-0 (hbk)
ISBN: 978-1-032-79905-6 (pbk)
ISBN: 978-1-003-49441-6 (ebk)

DOI: 10.4324/9781003494416

Typeset in Times New Roman
by Apex CoVantage, LLC

Contents

Figures and Tables

About the Editors and Contributors

Brizio, Adelina (Italy) specializes in cognitive science and has a research grant in the Department of Chemistry, Università di Torino. Her research interests are principally focused on social innovation and new technologies, workers' inclusion and communication, and public engagement on environmental issues.

Farrell J., Alysha (Canada) is Associate Professor in the Faculty of Education at Brandon University. Using arts-based methods like playwriting and narrative photography, she collaborates with others to tell stories about teaching in a warming world. Her research-art exhibition (2022) at the Art Gallery of Southwestern Manitoba called, Before I Go to Bed Tonight, featured the work of 17 young artists from southwestern Manitoba who reflected on the personal and collective impacts of the climate crisis.

Steele, Astrid (Canada) is a (retired) professor of science and environmental education within the Schulich School of Education at Nipissing University in North Bay, Canada. She holds a Ph.D. in Education and a Master of Education, both from the University of Toronto, and B.Ed., B.P.H.E., and B.A. from the Queen's University. She has a strong background in outdoor environmental education and secondary science education; her current interests are in teacher development in science and environmental education, and in international teaching experiences. Her research focuses on teacher development in STSE/STEM education, international collaborations, and the ramifications of climate change. She has been involved in various collaborative projects with colleagues in art, science, and international education.

Juen, Barbara is a clinical and health psychologist and professor at the University of Innsbruck, Institute for Psychology (Austria). She is the leader of the working group for emergency psychology with a research focus on clinical psychology, crisis intervention, and psychotraumatology. She is Head of Psychosocial Support at the Austrian Red Cross, and Scientific Advisor of the European Network for Psychosocial Support (ENPS) and the IFRC Reference Centre for Psychosocial Support.

Hickman, Caroline (United Kingdom) is a psychotherapist and lecturer at the University of Bath researching children and young people's emotional responses to climate change internationally for 10 years examining eco-anxiety and distress, eco-empathy, trauma, moral injury, and the impact of climate anxiety on relationships. She is co-lead author of a 2021 quantitative global study into 10,000 children and young people's emotions and thoughts about climate change. She has been developing a range of psycho-educational approaches to ecological distress including a psychological assessment model for eco-anxiety.

Herba M., Catherine (Québec, Canada) is Associate professor in the Psychology Department at the Université du Québec à Montréal (UQAM) as well as Researcher at the CHU Sainte-Justine Research Center. Her research programme aims to gain a better understanding of the intergenerational transfer of risk, and she has a particular interest in the longitudinal associations between parental mental health difficulties in relation to child development. She is interested in mechanisms underlying associations between parental mental health and child development and conducts her research within the context of longitudinal and cross-sectional studies to examine both risk and protective factors.

Malboeuf-Hurtubise, Catherine (Québec, Canada) is an associate professor at Bishop's University and holds an FRQS Young Scholar Junior 1 award. She is a licensed child psychologist and specializes in youth mental health. She is a leading figure in mindfulness and youth research, in art therapy and philosophy for children in the province of Quebec, Canada. She is a PI or co-PI on multiple grants, that all fund projects on children mental health and well-being in school settings.

Petersen, Christie (Canada) is a Ph.D. candidate in the Faculty of Education at the University of Manitoba. She is also currently an educator in the Winnipeg public school system. Her research interests include the study of early career teachers, teacher identity, and becoming teacher. She engages with philosophy, phenomenology, feminist theory, poststructuralism, and affect theory to theorize and critically take up the experiences of early career teachers. Her recent work considers becoming teacher with climate change.

Acquadro Maran, Daniela (Italy), holds a Ph.D. in work and organizational psychology, is Associate Professor at the Department of Psychology, Università di Torino. Her research interests are principally focused on eco-anxiety and pro-environmental behaviours, violent behaviour in the workplace and workers' inclusion and participation in organizational processes.

Wiseman, Dawn (Québec, Canada) is Associate Professor in the School of Education at Bishop's University in Ktinékétolékouac. She has engaged in thinking about Science, Technology, Engineering, Arts, and Mathematics

(STEAM) with young people and educators for over three decades, most often alongside Indigenous people, peoples, and communities in what is currently Canada. Her research examines how Western and Indigenous ways of knowing, being, and doing might circulate together in STEAM education, student-directed STEAM inquiry, the distinctiveness of Canadian science education research, and the possibilities of teaching and learning within the context of human-driven climate change.

Cognet, Emma (Quebec, Canada) is completing an individualized, interdisciplinary master's degree at Bishop's University. Her work brings together psychology and program evaluation on community-based quality employment programming for youth. Methodologically, she draws on narrative inquiry and storytelling to raise awareness of the ways educational policies influence how youth navigate interconnected public-facing institutions.

Sciberras, Emma (Australia) is Associate Professor and Clinical Psychologist at Deakin University. She holds leadership roles in the Deakin Child Study Centre and Centre for Social and Early Emotional Development. She is Honorary Research Fellow and Team Leader in the Health Services Research at the Murdoch Children's Research Institute.

Madon, Erin (Canada) teaches secondary science for the Ottawa Carleton District School Board. Having started her career where she grew up in Thunder Bay, Ontario, she developed a love for nature and the outdoors that has carried through her 17 years of teaching across the province. She now resides in Ottawa with her family, and is an active member of OSEAN – the Ottawa South Eco Action Network.

Rueter Veiga, Josefina (Canada) is a researcher and educator. She has taught in Nunavut and Ontario, at the elementary and post-secondary levels. Most recently, she completed her Ph.D. at York University focusing on climate change education. Her research interests include the intersections of Indigenous Science and Western Science, climate change and sustainability education, and outdoor education.

Fernando, Julian (Australia) is a lecturer in social psychology at the Deakin University. His work focuses primarily on understanding societal change and social action, particularly as it pertains to the natural environment and climate change. Within this broad area of interest, he has published research on several topics including environmentally sustainable utopian visions, climate anxiety and pro-environmentalism, veg*nism, and the sustainability of material artefacts.

Corkett K., Julie (Canada) is tenured Full Professor at Nipissing University's Schulich School of Education. Her research interests pertain to special education, educational psychology, teacher education, diversity and inclusion, pedagogy, and technology. She has published several peer-reviewed articles and book chapters pertaining to these topics and co-edited three books pertaining to microaggressions. She has presented her research at numerous international and national conferences and has been an invited guest lecturer both nationally and internationally.

Rune, Karina (Australia) is a psychology lecturer at the University of the Sunshine Coast and a practicing psychologist. She holds a committee position in the Australian Psychological Society College of Clinical Psychologists (QLD Section) and is a member of the International Society of Schema Therapy. Her research focuses on health and environmental psychology, specifically psycho-oncology and eco-anxiety. She explores psychological factors impacting the well-being and quality of life in individuals diagnosed with cancer and the implications of eco-anxiety on mental health, resilience, and coping.

O'Callaghan, Kirsty (Australia) commenced her Ph.D. candidature at the University of the Sunshine Coast in 2022 after completing a Bachelor of Communication (Honours 1) in 2021. Her research focuses on the role of gender in effective climate change communication. Over the past five years, she has supported the "Real Rural Women's Leadership" research projects as Research Assistant and has held numerous student leadership positions in the School of Law and Society. Before embarking on her Ph.D. research, Kirsty spent two decades as a consultant and continues to be active in her consulting business and several voluntary and pro bono roles.

McLarnon, Mitchell (Québec, Canada) is Assistant Professor in the Department of Education at Concordia University. His research interests include, but are not limited to, institutional ethnography, community-based and participatory research, visual methodologies, land-based/environmental education, community gardening, social and environmental justice, gentrification, food insecurity, and urban political ecology.

Pihkala, Panu (Finland) from the University of Helsinki is an expert in interdisciplinary eco-anxiety research. He hosts the podcast Climate Change and Happiness together with Dr. Thomas Doherty (https://climatechangeandhappiness.com/) and serves as an advisor for many projects about eco-anxiety. He has been awarded several prizes in Finland for his books about eco-emotions. When leading workshops, he often co-operates with artists and educators.

Nunn D., Patrick (Australia) is Professor of Geography at the University of the Sunshine Coast. His research has focused on climate change issues, understanding past and (likely) future human–climate interactions, and their implications for coastal livelihoods. This work has seen the publication of several books and more than 360 peer-reviewed publications. He has long been involved with the Intergovernmental Panel on Climate Change (IPCC);

he was Lead Author on the chapter about *Sea Level Change* in the IPCC's 5th Assessment Report and was Lead Author on the chapter about *Small Islands* for its 6th Assessment.

Esquivel L., Rebeca (Québec, Canada) is a science educator/communicator and Ph.D. student in the Department of Education at Concordia University. She is a Ph.D. student and her work uses participatory and storytelling methodologies to facilitate reflections around empowerment, connection, and activist practices. She is interested in issues of collective work in both formal and informal learning settings that open new possibilities for repairing material and other relationships.

Orman-Ditchfield, Rubie (Australia) is a secondary science teacher who has embarked on a Higher Degree by Research at the University of the Sunshine Coast. Her research focuses on understanding the psychological and emotional challenges that climate change poses to young people. Her research was inspired by her observations in the classroom, where she witnessed the profound emotional impact that learning about climate science had on her students. Her research aims to explore and develop educational strategies to support adolescents in managing their climate-related anxieties and fostering resilience in the face of environmental crises.

Grünthal, Satu (Finland/Lithuania) is Associate Professor of Finnish Literature at the Universities of Helsinki and Turku and holds the position of senior university lecturer at the University of Helsinki. Currently, she is working as a visiting associate professor of Finnish literature at the University of Vilnius, Lithuania.

Begotti, Tatiana (Italy) is a researcher in developmental and educational psychology at the University of Turin. Her research interests mainly concern risk behaviours in adolescence, promotion of well-being in different developmental contexts, peer relationships, bullying, and pro-environmental behaviours.

Léger-Goodes, Terra (Québec, Canada) is a Ph.D. student specializing in clinical psychology and eco-anxiety. Fuelled by a profound interest in the developing minds of children and the growing concern for environmental issues, her academic journey is driven by the desire to provide tools to children to learn to cope with the psychological impacts of climate change. Using creative methods, philosophy, children's literature, and photography, she strives to foster mental health, emotional literacy, and environmental awareness to create a sustainable and resilient future for the next generations.

Kelly, Tim is an adjunct fellow at the University of Canterbury, New Zealand. He specializes in environmental and educational psychology, with a particular focus on values and norms and how these relate to children's behaviour. He is Head of Science at Hurunui College, a small rural high school, and also teaches psychology online through NetNZ.

Lahtinen, Toni (Finland) is an associate professor and a senior lecturer at the University of Helsinki. He has conducted research on the descriptions of nature in Finnish literature. He has edited the first eco-critical anthologies in Finnish and is currently working on a monograph on literature in the age of Anthropocene. He is the coordinator and a founding member of the ENSCAN (The Ecocritical Network for Scandinavian Studies) research network. He is the principal investigator in the project Literature and reading in the era of climate crisis (Kone Foundation 2021–2025).

Kulcar, Vanessa is a psychologist at the working group for emergency psychology at the University of Innsbruck, Institute for Psychology (Austria). Her research focus is on the effects of global and systemic crises like the COVID-19 pandemic and climate change on the mental health and well-being of young people.

Abd-El-Aal, Wafaa Mohammed Moawad (Sultanate of Oman/Egypt) is Assistant Professor at College of Education at Sultan Qaboos University, Sultanate of Oman. She is also Associate Professor at Faculty of Education in Beni-Suef University, Egypt. She is interested in using technological, pedagogical, and content knowledge (TPACK) framework in science education. She is also interested in environmental education/citizenship as well as preparing teachers to teach science in inclusive schools.

Acknowledgements

We wish to acknowledge that this book was conceived and compiled on the territory of Nipissing First Nation, the territory of the Anishinabek, within the lands protected by the Robinson Huron Treaty of 1859. We are grateful to be able to live, learn, and grow on these lands with all of our relations.

1   Setting the Stage

Climate Change and Its Effect on Student Well-being

Julie K. Corkett and Astrid Steele

Today's youth are growing up in a world where climate change is not just some distant scientific theory but an indisputable and very tangible series of events worldwide. The year 2023 was marked as the hottest year on record with deadly heatwaves in Europe, North America, China, and Asia (Climate Adaptation Platform, 2024; Copernicus Climate Change Service, 2024). A three-year drought in the Horn of Africa that caused food insecurity for millions was followed by massive floods (World Weather Attribution, 2023). In Libya, more than 3,400 people died when three dams collapsed due to heavy rainfall (Climate Adaptation Platform, 2024; World Weather Attribution, 2023). Canada faced its most extreme wildfire season with more than 18 million hectares burnt, which not only caused people to flee their homes but the dense smoke also posed a significant health risk to people across Canada, the United States, and parts of Europe (Copernicus Atmosphere Monitoring Service, 2023; World Weather Attribution, 2023). Now in 2024, due to drought conditions caused by below-average rain and snowfall, Canada is battling hold-over fires that are once again causing mass evacuations (Taylor, 2024). Climate change has arrived, with no mercy.

Scientific consensus affirms that human activities, most notably the burning of fossil fuels (with the resultant greenhouse gases carbon dioxide, methane, and nitrous oxide), are destabilizing long-established global weather patterns (Brown et al., 2022; Intergovernmental Panel on Climate Change [IPCC], 2018), for example, increased jet stream winds (Shaw & Miyawaki, 2024). Deforestation, overfishing, invasive species, and numerous other human activities further contribute to Earth's faltering ecological systems. In fact, the potential collapse of the Atlantic Meridional Ocean Circulation (The National Aeronautics and Space Administration, 2023) would significantly alter the climate in Europe and raise global sea levels.

It is important to clarify that global warming should not be used interchangeably with climate change, as there is a clear and distinct separation between these terms. The 2018 IPCC Report defines climate change as the long-term alteration in Earth's climate, particularly in relation to changes in temperature, precipitation, and other atmospheric conditions. Although global warming is a single

DOI: 10.4324/9781003494416-1

contributing factor of climate change, it is often the focus of climate change discourse because the increase in greenhouse gases correlates with rises in global surface temperatures, escalating land-sea thermal contrasts and changing the trajectory of tropical storms and precipitation northwards (Monerie et al., 2022). Current modelling suggests that global warming will continue to increase reaching and very likely exceeding 1.5°C by 2040 (IPCC, 2023). A 1.5°C increase will result in catastrophic biodiversity loss, and a 2°C and 3°C increase will lead to irreversible extinctions of various plants, animals, insects, coral reefs, and loss of the Greenland and West Antarctic ice sheets (IPCC, 2023).

The grim impact of anthropogenic (human-caused) climate change is often revealed in media's doomsday rhetoric that paints a picture of catastrophe, urgency, irreversibility, and devastation. Influential figures, such as Greta Thunberg (2019), further amplify doomsday fears by urging people to panic, to feel daily fear, and to react as if their home were ablaze. With the relentless exposure to the dire consequences of climate change being inflicted on individuals, families, communities, and the world at large, it is not surprising that many people are grappling with climate anxiety (Albrecht, 2011; Clayton, 2020; Maran & Begotti, 2021; Ojala, 2015; Reyes et al., 2021). Today's students are not spared the experiences and fears of climate change and are acutely aware of the multifaceted impacts of climate change. Therefore, teachers may be growing increasingly concerned that classroom discussions examining the ramifications of climate change on both the local and global scales may affect the mental well-being of both students and educators through the manifestation of climate anxiety.

Research indicates that concern about climate change is prevalent among youth, young adults, and those who feel a strong connection to nature (Clayton & Karazsia, 2020; Ramadan et al., 2021; Pihkala, 2019; Sciberras & Fernando, 2022; Searle & Gow, 2010). Clayton and Karazsia (2020) found that up to 27% of Americans reported that their climate anxiety affected their ability to function, with younger people reporting a greater effect than older adults. Hyry (2019) examined the feelings climate change evoked in 2,070 citizens of Finland who were over the age of 15 and the effect of those feelings. Even though very few Finns reported having directly felt the effects of climate change, more than half of the participants (58%) reported that they were worried about climate change, with just over a quarter stating that "anxiety" accurately described their feelings. Similarly, the Australian Institute for Disaster Resilience's (2020) findings from a survey of 1,447 youth and young adults between the ages of 10 and 24 revealed that more than 80% of Australians over the age of 16 were concerned or extremely concerned about climate change. The fact that such a high proportion of youths are reporting that climate change is generating moderate or higher levels of anxiety suggests a need for educators to recognize and address climate anxiety in their classrooms.

It is important to recognize the distinction between eco-anxiety and climate anxiety. Both terms are often used interchangeably in reference to emotional and psychological distress experienced in response to climate change. However, a subtle distinction can be made between the two terms. Eco-anxiety is viewed as a broad term that encompasses a range of anxieties, including climate anxiety, as well as concerns associated with various environmental issues such as loss of biodiversity (Pihkala, 2020). Essentially, eco-anxiety acknowledges the collective impact that human activity has on the planet, and the potential consequences for ecosystems and future generations.

Climate anxiety focuses on the more direct experiences and consequences associated with a changing climate. Thus, climate anxiety refers to uncontrollable and chronic stress, worry, and fear about climate-induced endangerment to all forms of life to such a significant degree that it affects an individual's emotional, cognitive, and physical well-being (Dodds, 2021; Ramadan et al., 2021; Reyes et al., 2021). In essence, climate anxiety can be understood as an intense concern and fear about one's ability to survive amidst the challenges posed by climate change (Weintrobe, 2012). As climate anxiety is rooted in the concrete reality of the ongoing threat of climate change, it stands apart from generalized anxiety, which can be linked to improbable events (American Psychiatric Association, 2013). Therefore, climate anxiety can be considered the manifestation of the recognition of the urgency and scale of climate change on one's environment, and its direct impact on one's ability to survive; for many, it has become an existential threat. It is evident, therefore, that students who experience the effects of climate change and/or think critically about the consequences of climate change have the potential to experience climate anxiety.

It is essential, however, that climate anxiety is not viewed as pathological. Climate anxiety is not a medically recognized disease, medical condition, or mental disorder. Climate anxiety is a rational response to the scientifically validated threats that climate change poses to the environment, humanity, and ecosystems. Teachers, therefore, can be perceived as frontline workers who are charged with the essential task of rationally educating children and youth about climate change, its causes, impacts, and possible solutions by fostering critical thinking and encouraging active participation in environmental initiatives. Through an understanding of climate anxiety, teachers can approach climate change education in a sensitive manner, balancing the dire messaging of climate catastrophe with messages of hope and empowerment. As teachers become cognizant of the emotional and psychological impacts of climate change, they can embrace the necessity of creating safe spaces for students to express their concerns and emotions. Through dialogue and critically scrutinizing academic and practical approaches to address the universal challenges associated with climate anxiety in schools, readers will discover that climate anxiety is a phenomenon that is often neglected in climate change education discourse. The purpose of this book is to provide the opportunity to reflect upon the importance and impact climate anxiety has on education praxis.

Organization of the Book

The theoretical framework underpinning this book is multifaceted, integrating insights from psychology, education, and environmental studies. The chapters are informed by recent research on climate anxiety, psychology, and pedagogical strategies that enhance resilience to climate anxiety and promote constructive engagement with climate change. The first section of this book, *Climate Anxiety Within Schools*, begins with an exploration of how students' awareness of climate change may affect their emotional well-being in the form of climate anxiety.

Section 1: Climate Anxiety Within Schools

We begin Section 1 with Daniela Acquardo Maran, Tatiana Begotti, and Adelina Brizio's exploration of how different age groups respond to climate change. While adults often exhibit cognitive dissonance, youth tend to react with anger and a desire for more information. However, youth are not homogenous; their reactions and attitudes towards climate change vary based on their developmental stage, cognitive abilities, and emotional capacities. The way youth engage in daily pro-environmental behaviours affects their perceived well-being and reflects broader concerns for future generations, developing countries, and nature. Chapter 2 focuses on how youth respond to climate change and explores educational tools that can promote pro-environmental behaviour.

Julian Fernado and Emma Scriberras' expand upon Acquardo Maran et al.'s discussion by exploring the relationship between climate worry and mental health. The authors provide the reader with a sound understanding of what climate anxiety is and how it contrasts with concern and worry. They then explore the prevalence of climate worry in youth. They identify factors that facilitate constructive engagement with climate change, such as fostering hope and efficacy, and framing climate change in an empowering manner. The chapter concludes with practical recommendations to help teachers channel their students' climate anxiety into positive action.

In Chapter 4, Vanessa Kulcar and Barbara Juen delve into the moral implications of climate change and its impact on the mental health of youth. They explore the psychological conflict that arises when students' moral ideals clash with their own behaviour and societal norms. An emphasis of the chapter is the importance of addressing moral dilemmas within educational contexts for the fostering of trust and agency.

Tim Kelly expands upon the moral implications of climate by exploring the double blind of conflicting values that contribute to climate anxiety in students. Tim discusses how societal norms that promote consumerism and individualism clash with the urgent need for sustainable behaviours, which leads to feelings of cognitive dissonance and anxiety. The chapter concludes with strategies for helping children and youth navigate their conflicting values through critical thinking and educational programmes that promote climate-friendly behaviour.

In Chapter 6, Karina Rune, Rubie Orman-Ditchfield, and Patrick D. Nunn argue that Australian students are keenly aware that their generation is highly vulnerable to impacts of climate change and are frustrated by the lack of policy change and action to address climate change. The authors discuss their qualitative research examining youth's experiences with climate change and its impact on their mental health and well-being. They identify themes related to eco-anxiety, copying strategies, and resilience. The chapter underscores the need for targeted support and climate change adaptation policies within educational settings.

The section concludes with Dawn Wiseman et al.'s argument that education and psychology are grounded in an ethic of caring and are thus well positioned to assist teachers and students who are grappling with the uncertainties of climate change. The chapter examines strategies that foster supportive classroom environments where students feel empowered to engage with climate issues and envision a sustainable future. Thus, their chapter serves as a segue into the exploration of concrete teaching strategies designed and implemented for addressing climate anxiety in the classroom.

Section 2: Addressing Climate Anxiety in the Classroom

The second section of the book describes educational strategies that foster resilience, critical thinking, and agency in students. The innovative and creative approaches for addressing climate anxiety in the classroom help students to engage constructively with climate change rather than succumbing to despair and hopelessness. Christie Petersen and Alysha J. Farrell begin this section with an introduction to solarpunk. Solarpunk is a pedagogical approach that creates space for counternarratives that challenge oppressive social constructs and taken-for-granted ways of living that generate climate change, and replace it with a focus on community, inclusivity, and sustainable living. Petersen and Farrell discuss how solarpunk drama can help students process their climate emotions and engage with the complexities of climate change in hopeful and creative ways. The chapter includes a play script to illustrate how educators can use drama to foster hope and resilience.

Chapter 9 continues the exploration of drama as a means for addressing students' emotions arising from climate change. Josefina Ruter Viaga explores a theatre-based climate change education programme for elementary students. She discusses how participatory and creative methods can enhance students' understanding of climate change and their sense of agency. The chapter highlights the potential of theatre to empower students by providing a supportive space for expressing and working through climate-related emotions.

Toni Lahtinen, Satu Grünthal, and Veli-Matti Värri shift the discussion to the role of children's and young adult literature in environmental education. They explore how reading environmental literature aligns with educational goals and supports students' understanding of climate issues. The chapter emphasizes the importance of integrating literary studies into environmental education to enhance students' engagement and literacy skills.

Erin Madon, in Chapter 11, reflects upon the evolution of climate education in Ontario, Canada, and shares her experiences as a science teacher. She discusses pedagogical strategies for easing climate anxiety and empowering students. The chapter provides practical insights and student feedback on effective approaches to climate change education.

Caroline Hickman's examines the impact of climate anxiety on children and youth through a depth psychology and eco-psychology lens. Caroline introduces the reader to psycho-educational approaches that integrate the arts, stories, theatre, and outdoor activities to help students navigate their emotional responses to climate change. The chapter highlights the importance of creative and relational strategies in addressing climate anxiety.

The book concludes with an acknowledgement of the complexity of climate change as both a scientific and deeply personal issue. Through the recognition of the need for holistic educational strategies that address cognitive, emotional, and ethical dimensions of climate change teachers will be better equipped to support their students' emotional journey through climate change.

Positionality Statement

Before delving into the book, we wish to acknowledge that the chapters within this book are written from a Global North perspective of climate change and climate anxiety. Given that emotions are culture-bound and climate events vary from region to region, it must be acknowledged that climate anxiety is a unique experience and cannot be generalized from the North to the South (Aruta & Guinto, 2022). However, as Hickman et al.'s (2021) study revealed regardless of culture, income, climate, climate vulnerability, and exposure to differing intensities of climate-related events, youth from both the Global North and the Global South are feeling anxious about their future within an ever-progressive climate crisis. As with the Global North, countries in the Global South, such as the Philippines, demand the development of effective educational policies and interventions to address climate change and climate anxiety (Aruta & Guinto, 2022). Thus, a growing number of studies are suggesting that the negative emotions arising from climate change "have relevance for mental health at a global level" (Ogunbode et al., 2023). Given that youth from around the world are experiencing the negative effects of climate anxiety, we encourage readers to reflect upon how the impact of climate anxiety and the recommended educational strategies revealed within this book can be adapted to suit their own unique positionality.

References

Albrecht, G. (2011). Chronic environmental change: Emerging "psychoterratic" syndromes. In *Climate change and human well-being* (pp. 43–56). Springer.

American Psychiatric Association. (2013). Anxiety disorders. In *Diagnostic and statistical manual of mental disorders* (5th ed.). https://doi.org/10.1176/appi.books.9780890425596.dsm05

Aruta, J. J. B. R., & Guinto, R. R. (2022). Climate anxiety in the Philippines: Current situation, potential pathways and ways forward. *The Journal of Climate Change and Health, 6*, 1–4. https://doi.org/10.1016/j.joclim.2022.100138

Australian Institute for Disaster Resilience. (2020). *Our world our say: National survey of children and young people on climate change and disaster risk.* www.aidr.org.au/media/7946/ourworldoursay-youth-surveyreport-2020.pdf

Brown, A. M., Bass, A. M., & Pickard, A. E. (2022). Anthropogenic-estuarine interactions cause disproportionate greenhouse gas production: A review of the evidence. *Marine Pollution Bulletin, 174*. https://doi.org/10.1016/j.marpolbul.2021.113240

Clayton, S. (2020). Climate anxiety: Psychological responses to climate change. *Journal of Anxiety Disorders, 74*(102263), 1–7. https://doi.org/10.1016/j.janxdis.2020.102263

Clayton, S., & Karazsia, B. (2020). Development and validation of a measure of climate anxiety. *Journal of Environmental Psychology, 69*.

Climate Adaptation Platform. (2024, January 16). Ten extreme climate events of 2023. *Climate Change Report.* https://climateadaptationplatform.com/ten-extreme-climate-events-of-2023/

Copernicus Atmosphere Monitoring Service. (2023, June 27). *Europe experiences significant transport of smoke from Canada wildfires.* https://atmosphere.copernicus.eu/europe-experiences-significant-transport-smoke-canada-wildfires

Copernicus Climate Change Service. (2024, February 29). *OBSERVER: 2023 – A year of unprecedented heat and climate extremes.* www.copernicus.eu/en/news/news/observer-2023-year-unprecedented-heat-and-climate-extremes

Dodds, J. (2021). The psychology of climate anxiety. *BJPsych Bulletin, 45*, 222–226. https://doi.org/10.1192/bjb.2021.18

Hickman, C., Marks, E., Pihkala, P., Clayton, S., Lewandowski, R. E., Mayall, E. E., Wray, B., Mellor, C., & van Susteren, L. (2021). Climate anxiety in children and young people and their beliefs about government responses to climate change: A global survey. *The Lancet Planetary Health, 5*(12), e863–873. https://doi.org/10.1016/S2542-5196(21)00278-3

Hyry, J. (2019). Kansalaiskysely ilmastonmuutoksen herättämistä tunteista ja niiden vaikutuksista kestäviin elämäntapoihin. *Kantar TSN OY.* Sitra https://media.sitra.fi/2019/08/21153439/ilmastotunteet-2019-kyselytutkimuksen-tulokset.pdf

Intergovernmental Panel on Climate Change. (2018). Annex I: Glossary. In V. Masson-Delmotte, P. Zhai, H.-O. Pörtner, D. Roberts, J. Skea, P. R. Shukla, A. Pirani, W. Moufouma-Okia, C. Péan, R. Pidcock, S. Connors, J. B. R. Matthews, Y. Chen, X. Zhou, M. I. Gomis, E. Lonnoy, T. Maycock, M. Tignor & T. Waterfield (Eds.), *Global warming of 1.5°C. An IPCC special report on the impacts of global warming of 1.5°C above pre-industrial levels and related global greenhouse gas emission pathways, in the context of strengthening the global response to the threat of climate change, sustainable development, and efforts to eradicate poverty.* In Press. www.ipcc.ch/sr15/chapter/glossary/

Intergovernmental Panel on Climate Change. (2023). Summary for policymakers. In Core Writing Team, H. Lee, & J. Romero (Eds.), *Climate change 2023: Synthesis report. Contribution of working groups I, II and III to the sixth assessment report of the intergovernmental panel on climate change* (pp. 1–34). IPCC. https://doi.org/1059327/IPCC/AR6-9789291691647.001

Maran, D. A., & Begotti, T. (2021). Media exposure to climate change, anxiety and efficacy beliefs in a sample of Italian university students. *International Journal of Environmental Research and Public Health, 18*, 1–11. https://doi.org/10.3390/ijerph1879358

Monerie, P., Wilcox, L. J., Turner, A. G. (2022). Effects of anthropogenic aerosol and greenhouse gas emissions on Northern Hemisphere monsoon precipitation: Mechanisms and uncertainty. *Journal of Climate, 35*(8). https://doi.org/10.1175/JCLI-D-21-0412.1

National Aeronautics and Space Administration. (2023, June 5). *Slowdown of the motion of the ocean.* https://science.nasa.gov/earth/earth-atmosphere/slowdown-of-the-motion-of-the-ocean/

Ogunbode, C. A., Pallesen, S., Böhm, G., Doran, R., Bhullar, N., Aquino, S., Marot, T., Schermer, J. A., Wlodarczyk, A., Lu, S., Jiang, F., Salmela-Aro, K., Hanss, D., Maran, D. A., Ardi, R., Chegeni, R., Tahir, H., Ghanbarian, E., Park, J., . . . Lomas, M. J. (2023). Negative emotions about climate change are related to insomnia symptoms and mental health: Cross-sectional evidence from 25 countries. *Current Psychology, 42*, 845–854. https://doi.org/10.1007/s12144-021-01385-4

Ojala, M. (2015). Hope in the face of climate change: Associations with environmental engagement and student perceptions of teachers' emotion communication style and future orientation. *The Journal of Environmental Education, 46*(3), 133–148. https://doi.org/10.1080/00958964.2015.1021662

Pihkala, P. (2019). *Climate anxiety.* MIELI Mental Health Finland. Helsinki. https://mieli.fi/wp-content/uploads/2021/12/mieli_climate_anxiety_30_10_2019.pdf

Pihkala, P. (2020). Anxiety and the ecological crisis: An analysis of eco-anxiety and climate anxiety. *Sustainability, 12*(19), 7836. https://doi.org/10.3390/su12197836

Ramadan, A. M. H., Randell, A., Lavoie, S., Gao, C. X., Manrique, P. C., Anderson, R., & Zbukvic, I. (2021). Understanding the evidence for climate change concerns, negative emotions and climate-related ill health in young people: A scoping review. *medRxiv*, 1–24. https://doi.org/10.1101/2021.09.27.21264151

Reyes, M. E. S., Carmen, B. P. B., Luminarias, M. E. P., Mangulabnan, S. A. N. B., & Ogunbode, C. A. (2021). An investigation into the relationship between climate anxiety and mental health among Gen Z Filipinos. *Current Psychology*, 1–9. https://doi.org/10.1007/s12144-021-02099-3

Sciberras, E., & Fernando, J. W. (2022). Climate change-related worry among Australian adolescents: An eight-year longitudinal study. *Child and Adolescent Mental Health, 27*(1), 22–29. https://doi.org/10.1111/camh.12521

Searle, K., & Gow, K. (2010). Do concerns about climate change lead to distress. *International Journal of Climate Change Strategies and Management, 2*(4), 362–379.

Shaw, T. A., & Miyawaki, O. (2024). Faster upper-level jet stream winds get faster under climate change. *Nature Climate Change, 14*, 61–67. https://doi.org/10.1038/s41558-023-01884-1

Taylor, G. (2024, May 10). Drought fuels wildfire concerns as Canada braces for another intense summer. *Yale Climate Connections.* https://yaleclimateconnections.org/2024/05/drought-fuels-wildfire-concerns-as-canada-braces-for-another-intense-summer/

Thunberg, G. (2019, January 25). "Our house is on fire": Greta Thunberg, 16, urges leaders to act on climate. *The Guardian.* www.theguardian.com/environment/2019/jan/25/our-house-is-on-fire-greta-thunberg16-urges-leaders-to-act-on-climate

Weintrobe, S. (2012). The difficult problem of anxiety in thinking about climate change. In S. Weintrobe (Ed.), *Engaging with climate change: Psychoanalytic and interdisciplinary perspectives* (pp. 33–47). Routledge.

World Weather Attribution. (2023, December 22). *Climate change fuelled extreme weather in 2023; Expect more records in 2024.* www.worldweatherattribution.org/climate-change-fuelled-extreme-weather-in-2023-expect-more-records-in-2024/

Part 1

Climate Anxiety Within Schools

The growing awareness of climate change can have profound effects on the emotional well-being of children and young people, making this an area of increasing concern for both researchers and educators. As youth become more attuned to the environmental crises facing our planet, they are experiencing a range of emotional responses, from climate worry to eco-anxiety. This section delves into these complex emotional landscapes, examining how students' awareness of climate change can impact their mental health and exploring the educational strategies that can support them through these challenges.

Part 1 begins with an exploration of the origins of the climate crisis and the Anthropocene era, as presented by Acquadro Maran, Begotti, and Brizio. The authors trace the rise of environmental education, particularly through the lenses of ecofeminist ethics and eco-pedagogy, illustrating how these frameworks have shaped the way we approach teaching about the environment. By understanding the ethical dimensions of climate change, educators are better equipped to foster pro-environmental behaviours in students, tailoring their approaches to the developmental stages and cognitive capacities of their learners.

Building on this foundation, Fernando and Sciberras, in Chapter 3, delve into the intricate relationship between climate worry and mental health. They distinguish between the various forms of climate-related distress, such as climate anxiety and worry, and explore how these emotions manifest differently among students. Their analysis reveals that while climate worry can lead to negative mental health outcomes, it can also serve as a catalyst for positive action when addressed constructively. The way climate change is framed in educational settings plays a crucial role in either empowering students or exacerbating their fears.

As the discussion deepens in Chapter 4, Kulcar and Juen introduce the moral dimensions of climate change, highlighting the psychological struggles that arise when young people confront the ethical implications of their actions – or inactions – in relation to the environment. They examine how students often experience moral injury when their behaviours fail to align with their environmental ideals, leading to feelings of guilt and disillusionment. The authors argue for the creation of educational environments where these moral conflicts

DOI: 10.4324/9781003494416-2

can be safely explored, allowing students to reconcile their ideals with the realities they face.

Kelly further extends the exploration of moralization in Chapter 5 by examining the phenomenon of cognitive dissonance among children and youth. As students navigate a world where societal norms often clash with the urgent need for sustainable behaviours, they frequently find themselves torn between their desire to act responsibly and the pressures to conform to less environmentally friendly practices. This internal conflict can lead to significant anxiety, which Kelly addresses by offering strategies for educators to help students navigate these contradictions, ultimately fostering a more cohesive sense of environmental responsibility.

In Chapter 6, the narrative then shifts to the specific context of Australia, where the impacts of climate change are becoming increasingly visible. Rune and her colleagues provide a case study account of how young Australians are coping with the threats posed by rising temperatures, sea levels, and extreme weather events. Australian students, acutely aware of their generation's vulnerability to climate change, often feel disempowered by the lack of meaningful policy responses. The authors emphasize the importance of educational initiatives that not only enhance climate literacy but also build emotional resilience, offering students a sense of agency and hope in the face of daunting challenges.

Concluding this section, Wiseman and her co-authors bridge the disciplines of education and psychology to address the overwhelming existential threats that climate change presents to both students and teachers. They argue that an educational approach grounded in an ethic of care is essential for helping young people navigate these challenges. By fostering caring relationships and creating supportive classroom environments, educators can help students move from a place of fear and paralysis to one of hope and resilience. The chapter introduces practical strategies, such as arts-based interventions and student-directed inquiry, that can transform classrooms into spaces of care and empowerment, where students are encouraged to engage deeply with climate issues and envision a sustainable future.

Throughout Part 1, the emotional responses of youth to climate change are not only acknowledged but also examined through multiple lenses – ethical, psychological, and educational. As the narrative unfolds, it becomes clear that while the challenges are significant, there are also profound opportunities for educators to support their students in navigating their emotions. Through thoughtful, compassionate teaching practices, we can help young people not only cope with the realities of climate change but also become active participants in creating a more sustainable world.

2 Experiences in Environmental Education With Young People

An Overview

Daniela Acquadro Maran, Tatiana Begotti, and Adelina Brizio

Positionality Statement

We are teachers and researchers in Italy who specialize in psychology. Our interest in education for pro-environmental behaviour stems from our diverse areas of research including individual development, participation in organizational life, and communication. In the area of environmental behaviour, we have conducted comprehensive studies at national level and in collaboration with international research institutes. In relation to youth and young adults, we have implemented innovative interactive educational methods, including the use of digital games. We are also investigating how participation in pro-environmental associations mitigates climate change-related stress, anxiety, and eco-worry. The chapter is framed by this perspective.

Introduction

In this chapter, we use the term "environmental issues" (EIs) in its broader sense to refer to a complex set of different, interrelated phenomena caused in whole or in part by human activities (Harper & Snowden, 2017). The human population has grown at an impressive rate over the last century: from 2 billion people in 1930 to almost 8 billion today, a doubling in the last 45 years. All human activities, including the extraction of resources, the use of energy sources and raw materials, the production of waste, and all types of pollution increased as a result. There are several alarming EIs that present and future generations must face the following:

- climate change, for example, warming is leading to ocean acidification;
- the loss of biodiversity, for example, the extinction of plant and animal species;
- environmental pollution, for example, exposure to air pollution and its impact on health;
- the use of resources, for example, water quality and availability and the resulting increase in food prices; and

DOI: 10.4324/9781003494416-3

- the deterioration of human quality of life and social degradation, for example, the unequal availability of energy.

The degradation of the environment and the deterioration of society affect everyone, but the poorest people are the most vulnerable on the planet and they will pay the higher price. Although vulnerable people are mentioned in economic and political debates, their problems remain at the bottom of the scale. While the demographic increase is blamed, consumption is seldom named. Underrating consumption contributes to foster the injustice of the "ecological debt," especially between North and South worldwide, which is linked to trade imbalances (Menton et al., 2020). The debt of low- and lower-middle-income countries has become an instrument of control, since they have the most important reserves of the biosphere, which feed the development of upper-middle- and high-income countries.

It is true that many economic assumptions continue to justify the current world system, in which speculation and the search for financial returns disregard the consequences for the environment and people. However, EIs are receiving more and more attention at political and economic levels, and media are reflecting this growing interest (Mavrodieva et al., 2019). Today's children and young people are an important target in tackling EIs. There are 1.2 billion adolescents (aged 10 to 19; 18% of the world's population – UNICEF, 2010), and they are also the group that will pay the highest price for environmental degradation in the coming decades. At the same time, children are particularly sensitized to EIs, probably because in modern society they grow up with climate change and its future predictions: they have an awareness of how it affects or will affect themselves and others (Martin et al., 2020, 2022). Due to this awareness, if they are provided with effective education and communication on environmental sustainability, they can be an important lever for change (Turcotte-Tremblay et al., 2023). To achieve the goal of an effective education and communication with younger generations, the first important step we must take is to change our image of young people.

The traditional view of youth portrays young people as victims or as a problematic part of society, as passive recipients of adult-driven policies. Instead, young people can be seen as powerful catalysts for community change, acting as resources and competent citizens in their communities (Makhoul et al., 2012). Young people who are exposed to pro-environmental knowledge can in turn advocate for more sustainable behaviours and lifestyles at home (Larsson et al., 2010; Žukauskienė et al., 2021). Young people have emerged as active decision makers and play a central role in decision making, making them a crucial group for promoting environmental behaviour change (e.g. de Leeuw et al., 2015). Educational interventions targeting young people have the potential to influence not only their behaviour but also that of those close to them (Stuhmcke, 2012; Bandura & Cherry, 2020).

The Birth of the Climate Crisis

Rachel Carson, a biologist and zoologist, was the initiator of the environmental movement in the United States. In "Silent Spring" (1962), she described the impact of human activity on ecosystems and encouraged a change in national policy on pesticides. However, it was not until the 1990s that EIs gained a high profile in public opinion. Today, the scenario has changed radically, and ecology and sustainability are at the top of the agenda in many developing countries. To better emphasize the extent of human impact on the environment, Crutzen and Stoermer (2021, see also Bińczyk, 2019) proposed a new name for the current geological era, calling it the "Anthropocene." According to the authors, Anthropocene began in 1784, when James Watt patented the steam engine, the symbol for the start of the industrial revolution. Anthropocene is the era in which the physical, chemical, and biological properties of the Earth are strongly influenced both locally and globally by the effects of human activities, in particular, by the increase in carbon dioxide (CO_2) and methane (CH_4) concentrations in the atmosphere.

According to Zalasiewicz et al. (2020), the Anthropocene is characterized by an impact on rocks and sediments that will remain visible for millions of years. As suggested by several authors such as Roka (2020) and Letcher (2021), evidence for this human impact is provided by the following:

- Nuclear explosions: for example, the Manhattan Project nuclear tests in New Mexico (1945), in which 2,421 nuclear devices were detonated. The tests and bombs deposit radioactive isotopes, unstable atoms in the environment that leave a radioactive footprint on the earth.
- Fossil fuels: From 1850 to today, the concentration of carbon dioxide in the atmosphere has risen to a record 400 parts per million. Global CO_2 emissions have left their mark on Antarctic ice, plants, sandstone sediments, fossil bones, and shells.
- The extinction of species of plants and animals.
- New materials: Aluminium, cement, and plastic are the most widely used materials in human society. About 500 million tonnes of the former have been produced in 150 years, and concrete has reached such a production rate that today there is one kilo per square metre of earth. In addition, 500 million tonnes are produced every year, so that for years there have been veritable plastic islands such as the Pacific garbage patch.
- The geological footprint: Mining, deforestation, urbanization, coastal erosion, and extensive agricultural activities are changing the geology and thus the stratification of rock sediments. The effects will be visible for millions of years to come.
- Fertilizers: Their massive use has excessively increased the phosphorus and nitrogen content in the soil. This will leave a visible chemical footprint for millennia to come.

- Global warming: The current geological epoch is experiencing the strongest climate change, which is also caused by man's environmentally harmful activities.

Anthropocenologists distinguish three main periods of time in the current era: The first extends from the beginning of the industrial revolution to the Second World War; the second begins in 1945 and it is called the "great acceleration" when pre-industrial institutions in Europe collapsed and the new international economic system of free exchange arose; the third phase began around the year 2000, and it was confirmed by the scientific community in 2001 with the third Intergovernmental Panel on Climate Change (IPCC; Foster, 2001) report.

As awareness of the impact of humans on the Earth grew, the first attempts were made to create governance systems to regulate the relationship between humans and the natural environment. Not all scientists agree with the idea that the Anthropocene began with the Industrial Revolution. Lewis and Maslin (2015), for example, suggest that the year 1610 should be regarded as the start date of this era. They postulate that the exchange of animal and plant species between the New and Old Worlds was a pivotal moment after which the Earth's biota became increasingly globally homogeneous, creating a new Pangaea and setting the Earth on a new evolutionary trajectory. This historical and ecological turning point, Lewis and Maslin remind us, was called "Orbis Spike" because the Eastern and Western hemispheres of humanity were brought together after more than 12,000 years of separation, creating a single global world economic system. It is also the beginning of the rise of the capitalist way of life and the scientific revolution.

From Ecofeminism to Eco-Pedagogy

In Great Britain since the late eighteenth century and in the United States after the Civil War, the development of industrialization, with its destructive consequences for the environment, gave a strong impetus to the study of nature, conservation movements, and the protection of animals. Women have always been the most numerous and active in these movements (Ling, 2014; Öztürk, 2020). Many of them combined the observation of nature with literary activity and activism, founding so-called ecofeminism, a movement emerging from protesting militarism and the destruction of the environment. Ecofeminism sees ecological and social problems as fundamentally linked. Recognizing an inseparable violence against women, children, and nature, ecofeminists make connections not only between sexism, speciesism, and the oppression of nature but also between other forms of social discrimination – racism, classism, heterosexism, and colonialism – as part of Western culture's aggression against nature. Ecofeminism examines the oppressive structures of the system and identifies three stages in the development of the "logic of domination":

1. alienation: belief in one's own identity, individualism, autonomy;
2. hierarchy: elevation of a single characteristic as the only one that can define the self; and
3. domination: justification of the subordination of other beings based on their supposed inferiority and lack of this unique quality.

Ecofeminism emphasises the interconnectedness of all forms of life and offers an ethical theory that is based on not only separation or abstract individualism but also the values of inclusion, relationships, promoting the preservation of life, starting from the awareness of everyone's vulnerability. The contribution of feminist thought is fundamental to eco-pedagogy. Eco-pedagogy emerged in the twentieth century in the Latin American context, inspired by Freire's critical pedagogy (Kahn, 2010), with the aim of training individuals to understand the contribution of each living being and grasp their interconnections and interdependencies.

Both eco-pedagogy and ecofeminism aim to address the problems related to EIs (Begum, 2022). Ecofeminist ethics is primarily based on not only empathy, the ability to feel and listen, but also an emotional and intellectual practice, an ethics of compassion that embraces all living things. Feeling the connection with all the living things and nature requires intense attention to the reality of the other, as well as concentration and judgement, the ability to grasp the experiences of others. As naturalists, ecofeminists spread a genuine ecological awareness, an ethic of care for the environment, and show, for example, how children's literature can play a crucial role in ecology. An ecofeminist perspective on environmental literature for children can look for new ways through which these narratives can provide an antidote to the logical domination rooted in alienation and the myth of the separate self, juxtaposed with the logical new narratives of connection, interdependence and interplay among humans, animals, and nature. Since the 1990s, large parts of ecofeminism have been featured as a pedagogy that could oppose a notion of humanity as separate and in opposition to nature and reevaluate all forms of knowledge. In 1988, Ynestra King defined this ecofeminist project as rational re-enchantment.

Inspired by the concept of rational re-enchantment, Kurth-Schai (1992), ecofeminist, scientist, and educator, called for a radical change in the education system that promotes the development of perceptual and intuitive skills. To challenge the belief that cognitive activity is a unique characteristic, education should focus on a reconceptualization of human and non-human nature that experiences continuity, overcomes the artificial barriers between the living, and promotes a gift economy, an economy that relies on giving and receiving in contrast to the market economy, which is based on acquisition and retention. This process of transformation, which must begin with a person's self, relationships, community, and places, has also been elaborated and suggested in children's literature.

Many authors (see, for example, Ritchie, 2015; York, 2014; Lin et al., 2023) have created narratives that focus on the interdependence of humans, animals, and the natural world, and on the values of justice, sensitivity, reciprocity, and care. The stories address the ability of children and young people to act and influence the environment; they are narratives that promote a deep connection within oneself and nature. As Plumwood (2002) has argued, it is important to foster a new sensitivity to the local dimension in order to move towards an environmental ethic that fully recognizes the non-human sphere and recognizes the importance of all EIs.

The Environmental Education

Education is a key factor in realizing a sustainable future in which nature is preserved and people are treasured. Environmental education specifically aims to promote a change in attitudes and behaviours, encourage the adoption of pro-environmental behaviour (PEB), that is, a person's conscious action to minimize the impact on the natural world in the name of environmental protection (Jensen, 2002) and develop a sense of ethical and social responsibility (Mello et al., 2021). The birth of international environmental education in 1972 (see Palmer, 2002) is the year in which the Stockholm Conference took place. However, the first international document to include the concept of environmental education dates back to 1965, when the Bangkok Conference on the Conservation of Nature and Natural Resources referred to education as an essential tool for promoting the protection and conservation of cultural heritage. In 1975, the United Nations Educational, Scientific and Cultural Organization (UNESCO) and the United Nations Environment Programme (UNEP) launched an international programme to exchange experiences, ideas, and educational programmes and to carry out research and training in the field of environmental education at an international level. In the same year, the Belgrade Charter of 1975 defined the goals and objectives of environmental education as educating the world's population to be aware of and concerned about EIs; to lay the foundation for a population that has the knowledge, skills, spirit, motivation, and individual and collective sense of duty to solve current environmental problems and to prevent new ones (UNESCO-UNEP, 1975). In 1975, the Tbilisi conference emphasized the importance of environmental education in all age groups and at all levels of formal and informal education. Furthermore, environmental education must be global and be aware of changes in a rapidly changing world. In 1992, the United Nations Conference on Environment and Development reaffirmed the importance of education, particularly in promoting sustainable development and improving people's ability to address EIs (ONU, 1992, see also 2002). In 1997, the UNESCO International Conference in Thessaloniki emphasized the importance of environmental education as a tool for the exercise of personal and responsible choices (UNESCO, 1997). In 2002, the United Nations convened a World Summit on Sustainable Development in Johannesburg. On this occasion, environmental education was not explicitly addressed, but the importance of education accessible to all was recognized as a fundamental factor in the fight against underdevelopment. The year 2003 marked the beginning of a series of world congresses, the World Environmental Education Congress (WEEC), dedicated to environmental education with different experiences in schools and with local communities (Marcinkowski, 2009).

In the following ten years, numerous initiatives were taken, from Earth Day 2008 to Education for Sustainable Development Week 2012: food, agriculture, and ecosystems were the main themes. The goal was not to provide answers to specific questions but rather to stimulate critical thinking and awaken a sense of belonging and responsibility for the world in which we live. In 2009, the UNESCO World Conference on Education for Sustainable Development took place in Bonn, and education was considered a developmental factor for all peoples of the world, including the most disadvantaged.

In 2012, the United Nations Economic Commission for Europe drafted the document "Learning for the Future: Competences for Education for Sustainable Development," which recognizes that learning remains the basis for the development of a sustainable society; education is therefore intended not as an information tool but as a process that reforms the way one lives and perceives the environment (Madsen et al., 2015). In 2015, UNESCO launched the Global Action on Education for Sustainable Development (GAP) programme to align with the 2030 Agenda for Sustainable Development. The 2030 agenda signed in 2015 by the governments of the 193 member countries outlines the official Sustainable Development Goals to be tackled from 2016 with aims of achievement by 2030. The goals to be achieved relate to key development issues that balance the three dimensions of sustainable development (economic, social, and ecological) and aim to end poverty, fight inequality, tackle climate change, and build peaceful societies that respect human rights.

The following is a synthesis of the timeline of the various documents and initiatives on the emergence and development of environmental education (Table 2.1).

However, formal education on environmental topics is not without challenges (Ouariachi et al., 2019), especially when we target adolescents and young people. Young people are often considered the "interactive generation" for their tendency to use new and less formal communication. If we want to raise awareness of environmental and social challenges and prepare young people to actively participate in the search for appropriate solutions and their implementation, it is urgent to develop innovative, transversal educational programmes. These programmes need to incorporate multiple understandings and combine different learning and teaching approaches, including non-formal learning and digital technologies.

Table 2.1 The Timeline of the Documents and Initiatives on the Development of Environmental Education

Date	Event/Document
1965	Bangkok Conference on the Conservation of Nature and Natural Resources
1972	Stockholm Conference: The birth of international environmental education
1975	The United Nations Educational, Scientific and Cultural Organization (UNESCO) and the United Nations Environment Programme (UNEP) launched an international programme in the field of environmental education at an international level
	UNESCO and UNEP drafted the Belgrade Charter which defined the goals and objectives of environmental education
	The Tbilisi conference emphasized the importance of environmental education in all age groups and at all levels of formal and informal education
1992	In the United Nations Conference on Environment and Development, the ONU reaffirmed the importance of education in promoting sustainable development and improving people's ability to address environmental issues
1997	The UNESCO International Conference in Thessaloniki emphasized the importance of environmental education as a tool for the exercise of personal and responsible choices
2002	The United Nations in the World Summit on Sustainable Development in Johannesburg recognized the importance of education accessible to all as a fundamental factor in the fight against underdevelopment
2003	The World Environmental Education Congress (WEEC) was dedicated to environmental education with different experiences in schools and with local communities
2008	Earth Day
2009	The UNESCO World Conference on Education for Sustainable Development took place in Bonn and education was considered a developmental factor for all peoples of the world, including the most disadvantaged
2012	Education for Sustainable Development Week
	The United Nations Economic Commission for Europe drafted the document "Learning for the Future: Competences for Education for Sustainable Development," which recognizes that learning remains the basis for the development of a sustainable society
2015	UNESCO launched the Global Action on Education for Sustainable Development (GAP) programme to align with the 2030 Agenda for Sustainable Development

Perceived Negative Consequences Related to Environmental Issues and Environmental Education

While EIs (especially climate change) are a known threat to physical health, they also impact mental health and well-being as exposure to environmental content can elicit various, often negative, emotional responses, including anger, sadness, despair, fear, guilt, worry, and anxiety (Ojala et al., 2021). Cianconi et al. (2020) conducted an analysis of studies on the impact of EIs and climate change on mental health and distinguished three types of impacts: the direct impacts with immediate effects (e.g. heatwaves), indirect impacts in the short term with extreme events (e.g. hurricanes, floods), and indirect impacts in the long term (e.g. forced migration). All three forms of impacts can affect people's mental health and can therefore contribute to the occurrence of psychiatric illnesses (e.g. post-traumatic stress disorder, mood disorders such as depression, anxiety, increased suicide rates, substance abuse, and increased aggressive behaviour against themselves and others). Cianconi et al.'s (2020) study also emphasized that EIs could have significant consequences for the most vulnerable groups: children, the elderly, women, people with mental illness, and people on low incomes. In particular, young people's emotional responses to EIs can lead to a specific form of anxiety known as "climate anxiety" or "eco-anxiety" (e.g. Ogunbode et al., 2021, 2022). Indeed, some studies suggest that climate anxiety is more prevalent among young people than in older adults (Clayton & Karazsia, 2020; Hickman et al., 2021). As Boluda et al. (2022) state, eco-anxiety is characterized by negative feelings such as fear. According to philosopher Hans Jonas (1984), fear could serve as an incentive to search for strategies to adapt to the EIs. However, fear does not always stimulate action as it depends on the individual. Fear can potentially have a paralysing effect, or fear can manifest as excessive worry or anxiety, thereby affecting one's ability to search for adaptive strategies.

As Pinto and Grove-White (2020) note, school systems can implement environmental education programmes to support the mental health of children and adolescents. Environmental education can indeed provide a push towards pro-environmental behaviour (PEB) through the inclusion of activities such as contact with nature, which has been studied as an element of mitigating negative feelings related to EIs. Moreover, Smith and Sobel (2014) argue that teachers and educators should educate their students about the importance of environment before they present the facts of EIs. Furthermore, providing tools to understand EI (e.g. from the point of view of the impact of climate change on daily life, and in the long term) can help students understand what changes will occur. In addition to this understanding, it is also necessary to offer support in adapting to change, such as showing which behaviours can reduce the footprint of human activities on the environment. When young people feel more competent in dealing with environmental infrastructures, they tend to experience a higher level of self-efficacy. This creates a positive cycle in which actions aimed at reducing impacts are seen as beneficial, thereby increasing a sense of well-being. If, on the other hand, young people feel less competent, they may perceive their behaviour as ineffective, reinforcing feelings of uselessness and heightening the negative emotions of anxiety and fear.

Recognizing negative emotions such as fear and anxiety can help students understand how to deal with them. For example, group discussions among students can help normalize feelings of guilt, sadness, or anger. Recognizing that others share similar feelings can help with the identification of strategies to address one's feelings and build resilience. Baudon and Jachens (2021) point precisely to the importance of building resilience in coping with EI through reframing the problem and identifying positive ways to mitigate it. Teachers should, therefore, teach students to develop critical thinking skills (Nagel, 2005), such as the ability to research and identify reliable information, to understand what part of the pollution problem each individual can take on themselves (at an individual, social, and/or political level), and which groups can be targeted through public pressure to adopt solutions with lower environmental impact (Nagel, 2005). However, as Pihkala (2020) argues, educators should be careful not to overemphasize individual action as an antidote to climate change, as this could lead to burnout or give disproportionate importance to individual action over collective, governmental, and industry actions. Instead, individual action must be supported by collective action that gives value to individual action and integrates it into a broader action that affects the whole community.

Pro-Environmental Behaviour and Environmental Education

Several studies have found that negative emotions related to EIs are associated with PEB, which refers to actions that consciously attempt to minimize negative impacts on the environment (Kollmuss & Agyeman, 2002) or even help it (Steg & Vlek, 2009). Examples of PEB include minimizing resource and energy consumption and the use and disposal of toxic substances and waste (Whitburn, 2020). Brown and Kasser (2005) argue that participation in PEB increases personal well-being. Therefore, at a time when the Earth's climate is changing negatively and people's health and safety are at risk, it makes sense for people to participate in PEB. Studies on PEB and subjective well-being highlight the complexity of determining the relationship between the two variables. PEB is a product of both internal (psychological) and external (social, economic, physical, etc.) factors. Therefore, the concept of PEB has developed based on the role of different variables, such as the physical/natural environment in a person's identity (Clayton & Opotow, 2003); people's emotional attachment to the physical and natural environment (Frantz & Mayer, 2014), and the extent to which they view themselves as part of nature (Schultz, 2014); a combination of affective, cognitive, and experiential aspects (Nisbet et al., 2011); and attitudes/behavioural intentions and observable behaviour (Brugger et al., 2011). Bamberg and Moeser (2007) have proposed an integrative model in which PEB is driven by a combination of self-interest and pro-social motives. It should be noted that while people with a more anthropocentric orientation use PEB to protect the environment, they are motivated by the value of nature for improving people's quality of life: the values underlying anthropocentrism tend to be human-centred and utilitarian (Kopnina et al., 2018). From an environmental education perspective, it is important to understand how young people are guided by environmental values in order to determine the optimal content of their environmental education to foster their PEB. The positive relationship between subjective well-being and PEB is confirmed by other studies, such as that of Fang (2021), who found that in young people PEB leads to intrinsic motivation and triggers cheerful feelings.

As described earlier, environmental education supports PEB. An experimental study conducted by Duerden and Witt (2010) to assess the effects of direct (field) and indirect (classroom) experiences on the development of environmental knowledge, attitudes, and behaviours showed that indirect (or classroom) experiences were a significant, positive predictor of PEB. However, from their qualitative data, they also concluded that environmental education has a better effect when it is combined with direct experience (e.g. immersion in nature). In this case, the experience may influence attitudes towards environment and PEB. The lack of an educational programme (direct or indirect experience) could therefore negate the positive environmental values necessary for PEB. The American Pew Research Centre surveys, which argue for education as a means to promote PEB, have also shown that the lack of indirect or direct environmental experience is a barrier to PEB. Knowledge can be implemented through education because, as Stronge (2018) argues, education is an informative tool for behaviour change whose implementation costs are low and whose effects are long-lasting and strong, especially when the focus is on knowledge (cognitive domain) and behaviour (affective and psychomotor domain).

Some Examples of Environmental Education and Pro-Environmental Behaviours

Attention to PEB and its promotion has grown in recent years, and several examples of educational programmes aimed at young people can be found in the scientific literature. Overall, the relationship between environmental education and PEB is difficult to measure reliably. Aspects that influence the behaviours of people, especially young people, are many and their interplay is complex. It may be difficult to separate the effects of environmental education from other factors, both individual and contextual (Palupi & Sawitri, 2018).

Much research on this topic has been done in the Asian context, where economic growth and the rapid expansion of manufacturing industries since the 1960s have increased awareness of EIs and made governments more responsive to the importance of sustainable development and environmental protection through environmental education (He & Wu, 2020). Different frameworks of explanation and intervention to improve PEB have focused on the role of internal factors, such as knowledge, attitudes, values, motivation and individual norms, or external factors concerning collective and cultural issues, such as models, observation, social norms, and life experiences in nature.

The importance of knowledge about EIs is highlighted in the study by Zhang et al. (2022), conducted in China and involving 2964 junior high school students, with an average age of 12.70 years. Results stressed that EIs knowledge, mediated by environmental sense of responsibility, was a significant variable that can promote students' interest, attitudes, and engagement in educational projects related to environment health. Based on these results, Zhang et al. (ib.) suggested the importance of paying attention to both educational policymaking and educational promotion. Specifically, they suggested activities that can motivate proactive youth participation in school and strengthen collective environmental responsibility, such as the creation of handwritten newspapers on carbon neutrality. To this end, the authors presented two interesting examples: the China Science and Technology Museum, which organized the "30.60" Peak Carbon Emissions and Carbon Neutrality exhibition with interactive games and simulations on waste separation, and the Wisdom Valley in Wenyuhe Park in Beijing, a space full of interactive facilities with zero emissions, such as bicycles that can track the amount of carbon dioxide emissions saved by the cyclist.

In line with the contribution of Zhang et al. (2022), two similar studies were conducted by Djuwita and Benyamin (2019) and Raisya and Djuwita (2020) in Indonesia involving elementary students (aged between 10 and 13 years old) and junior high school students (aged between 11 and 15 years old), respectively. The data were collected through self-report scales, behavioural observations, and focus groups. The aim of these studies was to analyse whether students attending schools with nature-based curricula (Green schools) and schools with regular curricula showed a different level of implementation of PEB. In Indonesia, Green schools use the natural environment as a source of learning. In other words, students use various facilities available in nature and take the opportunity to learn natural phenomenon with direct experience. For example, they learn biology directly from gardening activities or using self-made compost for the plants in the school garden. The results stressed that for elementary children the sense of relatedness to nature was similar in the two groups, both appreciating and understanding their interconnectedness. In the case of both elementary and junior high students, those who attended a nature-based curriculum showed a significantly higher use of PEB compared to students attending a regular curriculum. Findings showed that students attending a regular curriculum tended to perform better in some specific dimensions related to PEB, such as ascription of responsibilities and personal norms. In other words, PEB of "green school" children did not seem to be based on knowledge or concern for environmental health but rather on a result of habituation derived from social modelling of friends.

These results suggest that a more general curriculum can promote aspects related to norms, values, knowledge, and beliefs, but these may not necessarily translate into behaviours. Students who participated in practical activities focused on environmental care demonstrated less theoretical reflection on EIs but were more likely to develop positive habits. Additionally, through observing the adoption of virtuous PEB, well-being was increased. Therefore, regardless of the type of curriculum, it is to better work on both the theoretical dimension, linked to knowledge and attitudes, and the experiential one, linked to actions, so that students develop not only a greater awareness of their own behaviour but also a better ability to translate knowledge and motivation into action.

The study by He and Wu (2020) aligns with previous research but places a stronger emphasis on various types of intervention. This research involved secondary schools in Taiwan and Singapore, selected based on their emphasis and productive learning outcomes in environmental education. Teachers and students were interviewed to understand the frameworks and strategies they used to promote environmental educations and their effectiveness. He and Wu identified four main paradigms, which can be traced back to Sauvé's typologies of environment (1996, see also Sauvè & Berryman, 2007) that characterize environment as nature (to be appreciated, respected and preserved), as a resource to be managed, as a problem to be solved, and as a community project in which to be involved. Activities proposed by the schools include, for example, cleaning up coastal areas, which gives students access to the natural environment in their immediate surroundings and gives them a better understanding of the habitat and its biodiversity. Other examples include setting up a composting area to teach students to reuse resources from the environment, building a specially designed solar-powered classroom that encourages students to explore other types of energy to replace current energy consumption, and finally, creatively using unwanted items to make student-made decorations. The latter activity, in particular, can build pride and a sense of community.

The results of He and Wu's study emphasized the strong interrelationship between external and internal factors in promoting PEB. On the one hand, external factors (especially promoted by governments) set clear goals for EIs and easily push people to follow certain norms for environmental protection. On the other hand, if these external factors place the relationship between the individual and the environment at the centre, people are more likely to develop an internal motivation to actively participate and maintain a greater interest in environmental protection.

The complex relationship between individual and contextual variables is also well highlighted by Nurwidodo et al. (2019), who collected data in schools adopting green programmes promoted by the Government in Indonesia. Through a qualitative design including direct and recorded observations, written sources and documents, and in-depth interviews, Nurwidodo et al. gathered information from principals, teachers, administrators, and students. The results suggested that the success of building PEB depends on active participation of the entire school community, including actions promoted by school leaders and external interventions, educational programmes oriented to reinforce individual skills, and the conditions of physical school settings. The integration of the different aspects acts both on individual knowledge and skills and on the construction of a context that reinforces the variables mentioned earlier.

Du et al. (2023) conducted another interesting study in a different geographical division of China, involving adolescent students (aged between 13 and 16 years old) who had participated in educational programmes on environmental protection. Based on the Theory of Planned Behaviour, the contribution of this study was to assess factors (both rational and affective) that may influence adolescents' willingness and attitudes to engage in PEB governance. The programme consisted in online sessions and field group practices geared towards students' development of problem-solving strategies. The findings showed that attitudes towards environmental protections and emotions (i.e. pro-nature emotions) significantly affected adolescents' PEB. From a didactic point of view, the authors suggested encouraging young people's interest and desire to learn more about EIs and help students feel that they have the skills, knowledge, and resources they need to adopt PEB, as these aspects play an important role in deepening nature-friendly feelings.

Finally, an interesting field experimental study conducted in India among school students aged 12–15 highlighted the role of incentive in promoting recycling plastic (Sherif, 2021). The results indicate that interventions based only on information did not promote change in the recycling behaviour of the students while providing incentives in addition to information encouraged an increase in recycling behaviour and a positive spillover from the incentive intervention to other PEB. Moreover, the spillovers were limited not to those students who increased their recycling behaviour but also to those who saw their peers responding positively to the intervention.

Overall, it is important to keep in mind that the studies reviewed did not always report the age of the students, the class attended, or describe in detail the characteristics of the intervention programmes. In most cases, the research evaluated the relationship between participation in a programme and the development of attitudes (or behaviours) oriented towards caring for the environment, but no specific information was reported on the activities carried out at school. Moreover, the aforementioned studies are almost all conducted in the Asian context, and they are difficult to generalize to contexts with consistently different environmental and cultural characteristics.

Conclusion

The aim of this chapter was to show how the perception of EIs influences the need to promote PEB. Ecofeminism was then presented as a perspective that inspired eco-pedagogy to educate young people to behave in an environmentally friendly way. Finally, empirical research on fostering PEB was presented and serves as useful food for thought for implementing programmes aimed at teaching PEB. The experiences of other research groups, highlighting the positive aspects and the inevitable limitations, can be a starting point for curriculum development. These studies also provide an opportunity to reflect on the need to share training experiences. There is a general call to investigate the phenomenon of PEB in longitudinal studies to better understand which factors (including eco-anxiety) experiences and contexts enable the formation of a stable PEB over time, that is, a behaviour that not only is adopted by an individual but can serve as an example for a social group. Longitudinal studies can hopefully clarify the nature of PEB and understand which variables remain stable (or are realized) over time.

References

Bamberg, S., & Möser, G. (2007). Twenty years after Hines, Hungerford, and Tomera: A new meta-analysis of psycho-social determinants of pro-environmental behaviour. *Journal of Environmental Psychology, 27*(1), 14–25. https://doi.org/10.1016/j.jenvp.2006.12.002

Bandura, A., & Cherry, L. (2020). Enlisting the power of youth for climate change. *American Psychologist, 75*(7), 945. https://doi.org/10.1037/amp0000512

Baudon, P., & Jachens, L. (2021). A scoping review of interventions for the treatment of eco-anxiety. *International Journal of Environmental Research and Public Health, 18*(18), 9636. https://doi.org/10.3390/ijerph18189636

Begum, M. S. (2022). Pertaining the feminist vision of ecocriticism for environmental justice against gender biases and women critics: A literature on the international and national perception. *International Journal of English Literature and Social Sciences, 7*(2), 182–186. https://doi.org/10.22161/ijels.72.23

Bińczyk, E. (2019). The most unique discussion of the 21st century? The debate on the Anthropocene pictured in seven points. *The Anthropocene Review, 6*(1–2), 3–18. https://doi.org/10.1177/2053019619848215

Boluda-Verdu, I., Senent-Valero, M., Casas-Escolano, M., Matijasevich, A., & Pastor-Valero, M. (2022). Fear for the future: Eco-anxiety and health implications, a systematic review. *Journal of Environmental Psychology, 84*, 101904. https://doi.org/10.1016/j.jenvp.2022.101904

Brown, K. W., & Kasser, T. (2005). Are psychological and ecological well-being compatible? The role of values, mindfulness, and lifestyle. *Social Indicators Research, 74*(2), 349–368. https://doi.org/10.1007/s11205-004-8207-8

Brügger, A., Kaiser, F. G., & Roczen, N. (2011). One for all? Connectedness to nature, inclusion of nature, environmental identity, and implicit association with nature. *European Psychologist, 16*(4), 324–333. https://doi.org/10.1027/1016-9040/a000032

Cianconi, P., Betrò, S., & Janiri, L. (2020). The impact of climate change on mental health: A systematic descriptive review. *Frontiers in Psychiatry, 11*, 74. https://doi.org/10.3389/fpsyt.2020.00074

Clayton, S., & Karazsia, B. T. (2020). Development and validation of a measure of climate change anxiety. *Journal of Environmental Psychology, 69*, 101434. https://doi.org/10.1016/j.jenvp.2020.101434

Clayton, S., & Opotow, S. (2003). Introduction: Identity and the natural environment. In S. Clayton & S. Opotow (Eds.), *Identity and the natural environment: The psychological significance of nature* (pp. 1–24). MIT Press.

Crutzen, P. J., & Stoermer, E. F. (2021). The "anthropocene" (2000). In S. Benner, G. Lax, P. J. Crutzen, U. Pöschl, J. Lelieveld & H. G. Brauch (Eds.), *Paul J. Crutzen and the anthropocene: A new epoch in earth's history. The anthropocene: Politik – economics – society – science* (Vol. 1, pp. 19–21). Springer. https://doi.org/10.1007/978-3-030-82202-6_2

De Leeuw, A., Valois, P., Ajzen, I., & Schmidt, P. (2015). Using the theory of planned behavior to identify key beliefs underlying pro-environmental behavior in high-school students: Implications for educational interventions. *Journal of Environmental Psychology, 42*, 128–138. https://doi.org/10.1016/j.jenvp.2015.03.005

Djuwita, R., & Benyamin, A. (2019). Teaching pro-environmental behavior: A challenge in Indonesian schools. *Psychological Research on Urban Society*, *2*(1), 26. https://doi.org/10.7454/proust.v2i1.48

Du, M., Chai, C. S., Di, W., & Wang, X. (2023). What affects adolescents' willingness to maintain climate change action participation: An extended theory of planned behavior to explore the evidence from China. *Journal of Cleaner Production*, *422*(4), 138589. https://doi.org/10.1016/j.jclepro.2023.138589

Duerden, M. D., & Witt, P. A. (2010). The impact of direct and indirect experiences on the development of environmental knowledge, attitudes, and behavior. *Journal of Environmental Psychology*, *30*(4), 379–392. https://doi.org/10.1016/j.jenvp.2010.03.007

Fang, S. C. (2021). The pro-environmental behavior patterns of college students adapting to climate change. *Journal of Baltic Science Education*, *20*(5), 700–715. https://doi.org/10.33225/jbse/21.20.700

Foster, B. (2001). IPCC third assessment report. *The Scientific Basis*. Geneva, Switzerland.

Frantz, C. M., & Mayer, F. S. (2014). The importance of connection to nature in assessing environmental education programs. *Studies in Educational Evaluation*, *41*, 85–89. https://doi.org/10.1016/j.stueduc.2013.10.001

Harper, C., & Snowden, M. (2017). *Environment and society: Human perspectives on environmental issues*. Routledge. https://doi.org/10.4324/9781315463254

He, J., & Wu, B. S. (2020). Pro-environmental behaviors in secondary schools: A comparison study of environmental education between Singapore and Taiwan. *Journal of Geographical Research*, *72*, 55–77. https://doi.org/10.6234/JGR.202011_(72).0003

Hickman, C., Marks, E., Pihkala, P., Clayton, S., Lewandowski, E. R., Mayall, E. E., Wray, B., Mellor, C., & van Susteren, L. (2021). Climate anxiety in children and young people and their beliefs about government responses to climate change: A global survey. *The Lancet. Planetary Health*, *5*(12), 863–873.

Jensen, B. B. (2002). Knowledge, action and pro-environmental behavior. *Environmental Education Research*, *8*, 325–224. https://doi.org/10.1080/13504620220145474

Jonas, H. (1984). *The imperative of responsibility: In search of an ethics for the technological age*. University of Chicago Press.

Kahn, R. V. (2010). *Critical pedagogy, ecoliteracy, & planetary crisis: The ecopedagogy movement* (Vol. 359). Peter Lang.

Kollmuss, A., & Agyeman, J. (2002). Mind the gap: Why do people act environmentally and what are the barriers to pro-environmental behavior? *Environmental Education Research*, *8*(3), 239–260. https://doi.org/10.1080/13504620220145401

Kopnina, H., Washington, H., Taylor, B., & Piccolo, J. (2018). Anthropocentrism: More than just a misunderstood problem. *Journal of Agricultural and Environmental Ethics*, *31*(1), 109–127. https://doi.org/10.1007/S10806-018-9711-1

Kurth-Schai, R. (1992). Ecology and equity: Toward the rational reenchantment of schools and society. *Educational Theory*, *42*(2), 147–163. https://doi.org/10.1111/j.1741-5446.1992.00147.x

Larsson, B., Andersson, M., & Osbeck, C. (2010). Bringing environmentalism home: Children's influence on family consumption in the Nordic countries and beyond. *Childhood*, *17*(1), 129–147. https://doi.org/10.1177/0907568209351554

Letcher, T. M. (2021). Why discuss the impacts of climate change? In T. Letcher (Ed.), *The impacts of climate change* (pp. 3–17). Elsevier. https://doi.org/10.1016/B978-0-12-822373-4.00020-3

Lewis, S. L., & Maslin, M. A. (2015). A transparent framework for defining the Anthropocene Epoch. *The Anthropocene Review*, *2*(2), 128–146. https://doi.org/10.1177/2053019615588792

Lin, J., Fiore, A., Sorensen, E., Gomes, V., Haavik, J., Malik, M., Mok, S.J., Scanlon, J., Wanjala, E., & Grigoryeva, A. (2023). Contemplative, holistic eco-justice pedagogies in higher education: From anthropocentrism to fostering deep love and respect for nature. *Teaching in Higher Education, 28*(5), 953–968. https://doi.org/10.1080/1356 2517.2023.2197109

Ling, C. (2014). The background and theoretical origin of ecofeminism. *Cross-Cultural Communication, 10*(4), 104–108. https://doi.org/10.3968/4916.

Madsen, K. D., Nordin, L. L., & Simovska, V. (2015). Linking health education and sustainability education in schools: Local transformations of international policy. *Schools for Health and Sustainability: Theory, Research and Practice, 81*, 109. https://doi.org/10.1007/978-94-017-9171-7_5

Makhoul, J., Alameddine, M., & Afifi, R. A. (2012). "I felt that I was benefiting someone": Youth as agents of change in a refugee community project. *Health Education Research, 27*(5), 914–926. https://doi.org/10.1093/her/cyr011

Marcinkowski, T. J. (2009). Contemporary challenges and opportunities in environmental education: Where are we headed and what deserves our attention? *The Journal of Environmental Education, 41*(1), 34–54. https://doi.org/10.1080/00958960903210015

Martin, G., Reilly, K., Everitt, H., & Gilliland, J. A. (2022). The impact of climate change awareness on children's mental well-being and negative emotions–a scoping review. *Child and Adolescent Mental Health, 27*(1), 59–72. https://doi.org/10.1111/camh.12525

Martin, G., Reilly, K. C., & Gilliland, J. A. (2020). Impact of awareness and concerns of climate change on children's mental health: A scoping review protocol. *JBI evidence synthesis, 18*(3), 516–522. http://doi.org/10.11124/JBISRIR-D-19-00253

Mavrodieva, A. V., Rachman, O. K., Harahap, V. B., & Shaw, R. (2019). Role of social media as a soft power tool in raising public awareness and engagement in addressing climate change. *Climate, 7*(10), 122. https://doi.org/10.3390/cli7100122

Mello, G., Reuter, J., Dias, M. F., & Amorim, M. (2021). Promoting environmental education with escape room activities: Critical factors for implementation. In *European conference on games based learning* (pp. 535–XV). Academic Conferences International Limited. https://doi.org/10.34190/GBL.21.109

Menton, M., Larrea, C., Latorre, S., Martinez-Alier, J., Peck, M., Temper, L., & Walter, M. (2020). Environmental justice and the SDGs: From synergies to gaps and contradictions. *Sustainability Science, 15*, 1621–1636. https://doi.org/10.1007/s11625-020-00789-8

Nagel, M. (2005). Constructing apathy: How environmentalism and environmental education may be fostering "learned hopelessness" in children. *Australian Journal of Environmental Education, 21*, 71–80. https://doi.org/10.1017/S0814062600000963

Nisbet, E. K., Zelenski, J. M., & Murphy, S. A. (2011). Happiness is in our nature: Exploring nature relatedness as a contributor to subjective well-being. *Journal of Happiness Studies, 12*, 303–322. https://doi.org/10.1007/s10902-010-9197-7

Nurwidodo, N., Al Muhdar, M. H. I., Rohman, F., Iriani, D., Herlina, H., & Fausan, M. M. (2019). Building pro-environmental behavior among school community of Adiwiyata green school. *Jurnal Pendidikan Biologi Indonesia, 5*(1), 23–32. https://doi.org/10.22219/jpbi.v5i1.7233

Ogunbode, C. A., Doran, R., Hanss, D., Ojala, M., Salmela-Aro, K., van den Broek, K. L., Bhullar, N., Aquino S. D., Marot, T. Schermer, J. A., Wlodarczyk, A., Lu, S., Jiang, F., Acquadro Maran, D., Najafi, R., Park, J., Tsubakita, T., Tahir, H., Albzour, M., . . . Karasu, M. (2022). Climate anxiety, wellbeing and pro-environmental action: Correlates of negative emotional responses to climate change in 32 countries. *Journal of Environmental Psychology, 84*, 101887. https://doi.org/10.1016/j.jenvp.2022.101887

Ogunbode, C. A., Pallesen, S., Böhm, G., Doran, R., Bhullar, N., Aquino, S., Marot, T., Schermer, J. A., Wlodarczyk, A., Lu, S., Jiang, F., Salmela-Aro, K., Hanss, D., Acquadro Maran, D., Ardi, R., Chegeni, R., Tahir, H., Ghanbarian, E., Park, J., Tsubakita, T., . . . Lomas, M. J. (2021). Negative emotions about climate change are related to insomnia symptoms and mental health: Cross-sectional evidence from 25 countries. *Current Psychology*, 1–10. https://doi.org/10.1007/s12144-021-01385-4

Ojala, M., Cunsolo, A., Ogunbode, C. A., & Middleton, J. (2021). Anxiety, worry, and grief in a time of environmental and climate crisis: A narrative review. *Annual Review of Environment and Resources*, *46*, 35–58. https://doi.org/10.1146/annurev-environ-012220-022716

ONU. (1992, 2002). Conferences. *Environment and sustainable development*. www.un.org/en/conferences/environment

Ouariachi, T., Olvera-Lobo, M. D., Gutiérrez-Pérez, J., & Maibach, E. (2019). A framework for climate change engagement through video games. *Environmental Education Research*, *25*(5), 701–716. https://doi.org/10.1080/13504622.2018.1545156

Öztürk, Y. M. (2020). An overview of ecofeminism: Women, nature and hierarchies. *Journal of Academic Social Science Studies*, *81*(13), 705–714. https://doi.org/10.29228/JASSS.45458

Palmer, J. (2002). *Environmental education in the 21st century: Theory, practice, progress and promise*. Routledge.

Palupi, T., & Sawitri, D. R. (2018). The importance of pro-environmental behavior in adolescent. In *E3S web of conferences* (Vol. 31, p. 09031). EDP Sciences. https://doi.org/10.1051/e3sconf/20183109031

Pihkala, P. (2020). Eco-anxiety and environmental education. *Sustainability*, *12*(23), 10149. https://doi.org/10.3390/su122310149

Pinto, R. S., & Grove-White, S. (2020). From climate anxiety to resilient active citizenship: When primary schools, parents and environmental groups work together to catalyse Change. *FORUM*, *62*(2), 251–266. https://doi.org/10.15730/forum.2020.62.2.251

Plumwood, V. (2002). Ecological ethics from rights to recognition: Multiple spheres of justice for humans, animals and nature. In N. Low (Ed.), *Global ethics and environment* (pp. 188–212). Routledge. https://doi.org/10.4324/9780203015254

Rachel, C. (1962). *Silent Spring*. Penguin Books.

Raisya, H., & Djuwita, R. (2020). Does green curriculum have an impact on pro-environmental behavior? A comparative study with middle schools. In *Proceedings of the 3rd international conference on psychology in health, educational, social, and organizational settings (ICP-HESOS 2018): Improving mental health and harmony in global community* (pp. 431–440). www.scitepress.org/Papers/2018/85905/85905.pdf

Ritchie, J. (2015). Social, cultural, and ecological justice in the age the Anthropocene: A New Zealand early childhood care and education perspective. *Journal of Pedagogy*, *6*(2), 41–56. https://doi.org/10.1515/jped-2015-0012

Roka, K. (2020). Anthropocene and climate change. In W. Leal Filho, A. M. Azul, L. Brandli, P. G. Özuyar & T. Wall (Eds.), *Climate action. Encyclopedia of the UN sustainable development goals*. Springer. https://doi.org/10.1007/978-3-319-95885-9_26

Sauvé, L. (1996). Environmental education and sustainable development: A further appraisal. *Canadian Journal of Environmental Education*, *1*(1), 7–34. https://cjee.lakeheadu.ca/article/view/490

Sauvé, L., & Berryman, T. (2007). Three decades of international guidelines for environment-related education: A critical hermeneutic of the United Nations discourse. *Canadian Journal of Environmental Education*, *12*(1), 33–54. https://cjee.lakeheadu.ca/article/view/630

Schultz, P. W. (2014). Strategies for promoting proenvironmental behavior. *European Psychologist*, *19*(2), 107–117. https://doi.org/10.1027/1016-9040/a000163

Sherif, R. (2021, March 8). Are pro-environment behaviours substitutes or complements? Evidence from the field. *Working paper of the Max Planck institute for tax law and public finance no. 2021–03*. https://ssrn.com/abstract=3799970 or http://doi.org/10.2139/ssrn.3799970

Smith, G. A., & Sobel, D. (2014). *Place-and community-based education in schools.* Routledge.

Steg, L., & Vlek, C. (2009). Encouraging pro-environmental behaviour: An integrative review and research agenda. *Journal of Environmental Psychology, 29*(3), 309–317. https://doi.org/10.1016/j.jenvp.2008.10.004

Stronge, J. H. (2018). *Qualities of effective teachers.* Ascd. https://mnprek-3.wdfiles.com/local-files/teacher-effectiveness/Qualities%20of%20Eff%20Teachers%20-%20Stronge.pdf

Stuhmcke, S. M. (2012). *Children as change agents for sustainability: An action research case study in a kindergarten (Publication number 61005)* [Doctoral dissertation, Queensland University of Technology]. https://eprints.qut.edu.au/61005/

Turcotte-Tremblay, A. M., Fortier, G., Bélanger, R. E., Bacque-Dion, C., Gansaonré, R. J., Leatherdale, S. T., & Haddad, S. (2023). Climate anxiety is associated with self-efficacy and behavioural engagement amongst adolescents: A cross sectional analysis. *Research Square.* https://doi.org/10.21203/rs.3.rs-2720187/v1

UNESCO. (1975, 1997). *Documents.* https://unesdoc.unesco.org/ark:/48223/pf0000059759

UNICEF. (2010). *Progress for children: Achieving the MDGs with equity* (No. 9). UNICEF.

Whitburn, J. (2020). *Children's environmental psychology, behaviour and education and wellbeing: The role of connection to nature* [Doctoral dissertation, Open Access Te Herenga Waka-Victoria University of Wellington]. https://doi.org/10.26686/wgtn.17145590.v1

York, R. A. (2014). *Re-connecting with nature: Transformative environmental education through the arts* (Publication number 1674750655) [Doctoral dissertation, University of Toronto, Canada]. ProQuest Dissertations Publishing.

Zalasiewicz, J., Waters, C., & Williams, M. (2020). The anthropocene. In F. M. Gradstein, J. H. Ogg, M. D. Schmitz & G. M. Ogg (Eds.), *Geologic time scale 2020* (pp. 1257–1280). Elsevier. https://doi.org/10.1016/B978-0-12-824360-2.00031-0.

Zhang, J., Tong, Z., Ji, Z., Gong, Y., & Sun, Y. (2022). Effects of climate change knowledge on adolescents' attitudes and willingness to participate in carbon neutrality education. *International Journal of Environmental Research and Public Health, 19*(17), 10655. https://doi.org/10.3390/ijerph191710655

Žukauskienė, R., Truskauskaitė-Kunevičienė, I., Gabė, V., & Kaniušonytė, G. (2021). "My words matter": The role of adolescents in changing pro-environmental habits in the family. *Environment and Behavior, 53*(10), 1140–1162. https://doi.org/10.1177/0013916520953150

3 Climate Worry and Mental Health in Young People

Approaches to Facilitating Constructive Engagement

Julian Fernando and Emma Sciberras

Positionality Statements

As an academic with expertise in social and environmental psychology, and a second-generation immigrant born in Australia to Sri Lankan parents, Julian Fernando's perspective on climate change is shaped by both personal and professional experiences. Julian's research on social change processes and the motivation for climate-related social action is influenced by active participation in climate protests and regular donations to environmental organizations. As an Australian-born clinical psychologist specializing in children's mental health, Emma Sciberras perspective on climate change and its psychological impacts is grounded in her both clinical experience and research background in neurodevelopmental disorders. It is through these lenses that the chapter is written.

As the threat of climate change becomes more salient, concerns have been raised about the potential for widespread negative mental health effects of the climate crisis. These concerns are particularly relevant for young people, who have grown up in a world of intensifying impacts of climate change, and who will live to see the potentially most serious consequences. However, as evidence emerges regarding the mental health impacts of worry about climate change, a complex pattern has been revealed. While concern about the climate does seem to be associated with negative emotions and mental health outcomes that association has not been observed strongly or consistently. Climate worry has also been shown to have positive associations with motivating climate action and pro-environmental behaviour. Thus, young people's worry about the climate can lead down both constructive and unconstructive paths. In this context, it is important to understand ways of not only assisting young people to cope with the negative emotional and mental health effects of growing up in a world with a changing climate but also fostering that concern for greater engagement towards mitigating the threat. Here, we present a review of the existing literature linking climate worry to both mental health and climate action and discuss some of the ways that young people can be assisted to best cope with their concerns about the climate.

DOI: 10.4324/9781003494416-4

Climate Worry, Mental Health, and Climate Action

In recent years, rapidly growing research literature has emerged attempting to understand the mental health impacts of worry about climate change. This research attention is driven by concerns about a potential impending mental health crisis as more and more people face the prospect of a climate change-affected future, and studies suggest that large percentages of the population are worried about climate change (e.g. Leiserowitz et al., 2021). Despite the gloomy prognosis, however, the literature has evidenced a diversity of findings, showing a potentially complex relationship between climate worry and mental health, as well as associations between climate worry and action to mitigate climate change. For example, a recent review of the health implications of eco-anxiety by Boluda-Verdu and colleagues (2022) identified several studies showing an association between eco-anxiety and some form of mental health problem (e.g. depression, anxiety, psychological distress), but also a number of studies showing evidence of eco-anxiety being associated with climate activism and pro-environmental behaviour.

As the reader may already have noted, the terminology used to describe concern about climate change varies between studies (e.g. concern, worry, anxiety). These terms are sometimes used relatively interchangeably but may have important implications for the interpretation of study findings and the relationships observed with other variables such as broader mental health and pro-environmental action. In general, climate or eco-anxiety tends to be defined as a stronger form of concern than, for example, climate worry (Clayton & Karazsia, 2020; Pihkala, 2020). However, the definition and measurement of climate/eco-anxiety has varied between studies from an emotional state like fear or worry, to something more akin to pathological anxiety or anxiety disorder (Pihkala, 2020). An additional complication here is that these different states are often measured using simple questionnaire items (e.g. "How worried are you about climate change?") which partly index the connotations those terms have for participants. In reviewing the literature later, we have been faithful to the terminology used in those studies but note that these terms are often not clearly defined and that this remains a limitation of this field of research.

Several reviews and large, cross-national studies have shown that climate change is a source of worry and other negative emotions for many people (Boluda-Verdu et al., 2022; Clayton, 2020; Ogunbode et al., 2021). Furthermore, there is evidence that climate worry (and other concerns related to threats to the natural environment) is associated with higher rates of psychological distress, anxiety, and lower subjective well-being (Clayton & Karazia, 2020; Curll et al., 2022; Gago & Sa, 2021; Reyes et al., 2021; Schwartz et al., 2022). For example, a large, 25-country study (with participants from all inhabited continents) by Ogunbode et al. (2021) showed negative climate-related emotions to be associated with lower self-rated mental health and increased insomnia symptoms.

Since these associations have been demonstrated in large multi-country studies, we have evidence that climate worry has potential implications for mental health and well-being worldwide.

The strength and consistency of the climate worry–mental health association has, however, varied. Another study by Ogunbode and colleagues (2022) across 32 countries (including those from the Global North and South) showed that while 46.8% of the sample were "very" or "extremely" worried about climate change, these rates varied quite dramatically between countries. And while several studies have demonstrated some association between climate worry and psychological distress, others have shown only small effects of climate worry on well-being (e.g. Schmitt et al., 2018), or indeed no effect at all (e.g. McBride et al., 2021). A recent study of over 1,000 participants in the UK by Whitmarsh et al. (2022) also showed that while rates of climate *concern* were relatively high, levels of climate *anxiety* were quite low, suggesting that the more impairing, clinical forms of anxiety are not as prevalent as general worry.

As noted previously, this literature also points to an association between climate worry and participation in pro-environmental action and activism (Bouman et al., 2020; Gregersen et al., 2021; Ogunbode et al., 2022; Smith & Leiserowitz, 2014; Verplanken et al., 2020). A study by Sangervo and colleagues (2022) also demonstrated that a majority of participants experiencing climate anxiety believed that it increased their pro-environmental behaviours. Thus, it may be, at least in some populations, that not only is climate worry associated with pro-environmental action but people regard climate worry as something which motivates them to act. In addition, studies have shown the capacity for participation in climate action to influence the effects of climate worry on broader mental health. For example, Curll et al. (2022) found that individual climate action reduced the psychological distress caused by climate worry, while collective climate action increased psychological distress. Schwartz et al. (2022), on the other hand, showed engagement in collective climate action moderated the association between climate anxiety and depression symptomology such that this association was non-significant for those with higher engagement in action.

This collection of findings is consistent with the notion – noted repeatedly among researchers and theorists in the area – that worry or concern about climate change represents a rational response to a serious threat, and so should not necessarily be interpreted as pathological or reflecting, for example, clinical forms of anxiety (see Brophy et al., 2023 for a review). Furthermore, given the evidence linking climate worry to engagement in climate change mitigation-related activities, some level of worry may even contribute to positive activities to alleviate the threat (and potentially affect broader mental health outcomes). In our previous work (Sciberras & Fernando, 2022), we have referred to this as the capacity for people to have *constructive* or *unconstructive* responses to climate worry. This distinction draws on the literature on worry (McNeill & Dunlop, 2016) which suggest that constructive forms of worry can lead to preparedness to deal with threats, whereas unconstructive worry predicts more negative outcomes such as generalized anxiety or depression. At present, however, this set of relationships is complex, and the factors which predict following a constructive or unconstructive pathway are unclear.

Climate Worry Among Young People

While the research referenced earlier is instructive as to the likely outcomes and associations of climate worry among young people, many existing studies have been conducted with adult participants. It is commonly hypothesized that younger people will be more susceptible to climate worry because (1) they have grown up in a world in which climate change has always been a salient concern, and (2) they are more likely to see the most serious potential consequences of climate change within their lifetimes. In addition to these factors, younger people are typically limited in their capacity to influence political solutions to climate change (by, for example, being unable to vote), and so may therefore experience a sense of powerlessness which may contribute to climate worry and other negative emotions.

Consistent with these notions, a study of 10, 000 young people (aged 16–25) across multiple countries[1] has shown that a substantial percentage of young people in several countries report climate worry and other negative emotions and pessimistic beliefs, as well as feelings of betrayal by adults and governments (Hickman et al., 2021). In that study, more than 40% of respondents in all countries surveyed reported feeling "very" or "extremely" worried about climate change. Indeed, there is also evidence that rates of climate worry and other negative emotions are higher and/or increasing faster among young people (Swim et al., 2022), or have increased over time (Ojala, 2023). Nevertheless, consistent with the mixed findings described for the general population, overall rates of high climate-related worry or distress among young people appear relatively low in some studies. For example, Vercammen et al. (2023), in a study of 16- to 24-year-olds in the UK, found only 10% to be classified as experiencing high climate distress. Similarly, in our own work on the topic (Sciberras & Fernando, 2022), we conducted a longitudinal study of over 2,000 young Australians, assessing climate worry at four time points across eight years from ages 10–11 to 18–19. We found a number of different patterns of climate worry over time. While approximately 37% were experiencing persistently high or increasing climate worry, the remainder were experiencing moderate (25%), low (17%), or even decreasing worry (13%) over time. Furthermore, we showed that while those exhibiting persistently high worry did show higher levels of depression symptoms, those experiencing increasing worry (with comparable levels to the persistently high group at the last time point) did not show greater depression symptoms than those with moderate worry. A similar study by Veijonaho et al. (2023) found four profiles among young Finnish people, showing various patterns of climate distress, climate denial, and pro-environmental behaviour. These studies indicate significant variation in the way young people respond to climate change, as well as variation in the relationship between climate worry and broader mental health issues.

Overall, the literature on climate worry among young people is not reflective of an epidemic of climate-related mental ill-health and suggests that the relationship between climate worry and broader negative mental health outcomes is not a simple one. In the sections which follow, we consider three issues which have likely contributed to the observed inconsistency in levels of climate worry in the population and the observed relationships between climate worry and broader mental health: (1) constructive and unconstructive responses to climate worry, (2) the direction of causality in the climate worry–mental health relationship, and (3) variation in how climate worry and mental health have been conceptualized and measured across studies.

Three Factors to Explain the Climate Worry–Mental Health Relationship

A first explanation for the complexity of the relationship between climate worry and broader mental health is, as we have already noted, that worry can contribute to readiness to act against a threat and thereby to constructive engagement with a problem (McNeill & Dunlop, 2016). This is consistent with the presence of young people at the forefront of global climate activism movements. A study by Ojala (2007) of young environmentalists and other activists showed that worry was viewed as a necessary emotion in response to threat, which could be coped with by balancing with positive emotions such as hope and by engaging collectively to address the problem. Thus, worry can be turned in a constructive direction and appears to be an important source of motivation for those young people at the forefront of climate activism.

A second consideration in understanding the complexity of the relationship between climate worry and mental health is that even where climate worry has been associated with broader negative mental health, the direction of causality is not always easy to understand. Given that this is an emerging field of research, many existing studies are cross-sectional (see Boluda-Verdu et al., 2022; Treble et al., 2023 for reviews), so it is not clear whether broader mental health issues are being driven by worry about climate change, or whether those who already experience conditions such as anxiety or depression are more likely to worry about climate change. Studies typically do not include underlying mental health as a possible explanation for associations between eco-anxiety and negative mental health, but some results suggest that this may be a factor. For example, Vercammen et al. (2023) showed one of the significant predictors of climate distress to be existing diagnosis/treatment for a mental health issue. In our work (Sciberras & Fernando, 2022), we also speculated that the high persistent climate worry group observed in our study may reflect a group of young people prone to worry generally (not just about climate change).

A similar question has been raised regarding the relationship between climate worry and climate activism. That is, does climate worry predict climate action or does engaging more in climate action (including, for instance, gaining more knowledge of the topic) facilitate further anxiety? For example, higher climate anxiety has been observed among young climate activists (aged 15–24) than among non-activists in Turkey (Ediz & Yanik, 2023). This suggests that climate anxiety may be a spur to action but perhaps also that greater knowledge of, and participation in, the climate cause, can facilitate further anxiety.

A third consideration, which is persistently in the background of this issue, is the problem of constructs and measurement of climate worry/anxiety/emotions. While we will not consider this factor at length (as it is of limited relevance in understanding how young people can be supported to deal with climate worry), it is important to note that some of the inconsistency in findings in the climate worry field of research may be traced to a lack of consistent conceptualization and measurement of responses to climate change (see Ramadan et al. (2023) on the variation in measurement in this field of research).

While climate change anxiety and eco-anxiety are the most commonly used terms to describe climate-related concern or worry (among lay people and academics), as detailed earlier, they are defined in various ways across studies (Pihkala, 2020; Treble et al., 2023) and possess connotations of clinical anxiety which are not the case for more neutral terms like climate "worry" or other emotional responses to climate change. We can speculate that when asked whether they experience "anxiety" about climate change, participants in studies may treat this as stronger or more serious compared to if they were asked if they are "worried" or "concerned." This speculation is supported by Whitmarsh et al. (2022) which showed higher levels of "concern" compared to "anxiety." Similarly, the measurement of these constructs varies (at times significantly) between studies. For example, a recent meta-analysis found a significant relationship between climate change anxiety and broader mental health concerns (Gago et al., 2024). However, this analysis contained studies using a specific measure of climate anxiety (Clayton & Karazia, 2020) designed to be similar to measures of clinical anxiety, which has been noted as assessing a stronger, clinical form of anxiety compared to general worry (Pihkala, 2020). It makes sense that a measure of more significant anxiety associated with climate change would be associated with broader mental health outcomes, but the same relationship may not pertain when using measures of constructs like climate "worry" or "concern." While this may be an issue of greater relevance for the research community, it is also important in accurately understanding the experience of climate change among young people. Depending upon how the question is asked (e.g. anxiety or worry), a very different picture may emerge regarding young people's level of concern and whether it is likely to have broader effects.

The preceding review of research pertaining to climate worry presents a complex picture of its relationships to general mental health and climate action. In the following section, we review two main approaches to identifying variables which may explain how climate worry can lead to constructive or unconstructive responses. Understanding these approaches can inform us as to how best to support young people to cope with climate worry and potentially channel it into engagement with climate change mitigation.

Coping Styles in Young People

The first approach, most prominently associated with the work of Ojala and colleagues, is to understand coping strategies as mediating or moderating the relationship between climate worry and (1) mental health and (2) pro-environmental engagement. Indeed, in a commentary to our paper (Sciberras & Fernando, 2022), Ojala (2022) urges the consideration of coping as a next step in explaining constructive and unconstructive responses to climate worry. Within this programme of research, Ojala and colleagues have examined three styles of coping: problem-focused, emotion-focused, and meaning-focused. Problem-focused coping involves attempting to solve the problem (e.g. taking action to mitigate climate change). Emotion-focused coping entails attempting to regulate the unpleasant emotion arising from the threat of climate change, while meaning-focused coping involves fostering positive thoughts and emotions (e.g. hope and optimism about the climate movement) rather than downregulating negative emotions. Across several studies (Ojala, 2012a, 2012b, 2013), Ojala and colleagues have shown meaning-focused coping to be positively associated with well-being, as well as to pro-environmental behaviour.

Recent research (Wullenkord & Ojala, 2023) has shown meaning-focused coping to potentially have effects on both climate pessimism and pro-environmental behaviours via different pathways. Meaning-focused coping (and in one study, optimism) buffered the effect of climate worry on climate pessimism. In addition, meaning-focused coping predicted pro-environmental behaviour, with that effect mediated by problem-focused coping. Thus, meaning-focused coping appears to have an important role to play in regulating negative feelings about climate change, as well as motivating climate change-mitigation activities. With this in mind, it is worthwhile examining the content of meaning-focused coping and how it might be implemented (see the following section on facilitating engagement).

An additional element of Wullenkord and Ojala's (2023) study, with implications for our second section on supporting young people to deal with climate change, is that two different types of climate worry were assessed: macro and micro. While micro worry (concern for oneself, friends, and family) was associated with climate pessimism, macro worry (concern for future generations, nature, etc.) was associated with problem-focused coping and pro-environmental behaviour. This suggests the importance of understanding the target of worry and suggests ways in which the issue of climate change may be framed to elicit more engagement and less pessimism.

Framing Climate Change to Enhance Empowerment

While much existing research and public discourse regarding negative feelings about climate change (especially that pertaining to the experiences of young people) has been framed in terms of climate- or eco-anxiety, emerging research indicates that we can gain a richer understanding of young people's reactions to climate change by looking at a greater range of emotional and motivational responses. Wang et al. (2018) framed this question around the notion of what it means to "care" about climate change. That is, are there different ways of thinking or feeling about climate change which may lead us to different emotional reactions (and potentially different mental health and motivational outcomes)? For example, the research by Hickman et al. (2021) cited earlier suggests that many young people feel a sense of *betrayal* given the inaction by older generations that has led to the current climate threat. This suggests that while anxiety may be one emotional response, young people may also experience feelings such as anger.

Several studies have demonstrated a range of possible emotional responses to climate change in addition to worry or anxiety (e.g. fear, sadness, grief, optimism, hope) (Galway & Field, 2023; Hickman et al., 2021; Verlie et al., 2021; Wang et al., 2018). Research has also demonstrated that various emotions can motivate climate action (see Harth, 2021, for a review), and emotion research has long identified different kinds of emotional response with characteristic action orientations (e.g. Frijda, 1993; Scherer, 1984). Stanley et al.'s (2021) work – albeit among adults – showed that while eco-anger predicted both better mental health outcomes and greater engagement in pro-climate behaviours, eco-anxiety and eco-depression were associated with poorer well-being and less engagement. This finding suggests that there are differences even among a set of relatively similar negative emotional responses, and that frustration or anger are more likely to lead to motivation (or what we have broadly characterized as "constructive" responses). This work highlights the capacity for understanding the range of emotional responses of young people, and potentially framing the issue of climate change in ways that encourage engagement and well-being, while avoiding framings which may be discouraging or demotivational. In addition, as the work on hope by Ojala (2012a, 2012b, 2015) cited earlier points out, emotional reactions to climate change need not only entail negative feelings, and incorporating positive emotions may also have a role to play in generating constructive responses.

Another form of framing that may be useful in this context is aligning the challenges of climate change with typical concerns of the developmental stage that children/young people are at: the "motive-alignment hypothesis," as suggested by Thomaes et al. (2023). For example, a key motive of adolescence is autonomy and peer connection/status, and so presenting engagement with climate change as achieving that motive (e.g. standing up against injustice, speaking out against those with authority, contributing to climate initiatives at school with peers) may not only elicit greater action from young people towards climate change mitigation but also do so in a way that is not perceived as overburdening (because it is achieved in a way that is consistent with existing motives).

A caveat at this point – following from the notion of overburdening young people – is that while there is some hope in the notion that climate anxiety can motivate climate action (and indeed, that climate action may help to alleviate climate anxiety), there is also concern about placing the responsibility to "solve" climate change on young people, and of individualizing the problem (see Brophy et al., 2023). Much existing research seems to implicitly present (poorer) mental health and greater engagement or activism as two potential pathways arising from climate change worry. There is an implication here that if one is worried about climate change, one is obliged to engage in activist or mitigation action, or that one should relieve one's worry through this kind of action. However, it has been noted that emphasizing individual responsibility for acting on climate change (and in the process, overcoming anxiety) can overburden young people, who may already feel that the burden has fallen on them due to the inaction of older generations (Hickman et al., 2021; Verlie et al., 2021).

Regulating Concerns About Climate Worry and Facilitating Engagement

It is important to recognize young people as being part of a broader support system and should not be expected to carry the burden of climate worry alone. Parents and teachers, key figures in a young person's life, play a vital role in understanding young people's feelings about climate change and helping to facilitate effective ways of coping. Therefore, many of the points we make later relate to the ways that adults can facilitate coping and engagement with young people, rather than solely focusing on young people themselves. A recent publication by Sanson et al. (2022) provides a commentary on the multiple ways that young people can be supported, both directly and indirectly via parents and other means (e.g. school education programmes). In addition, the body of research by Ojala provides practical guidance as to how positive coping may be facilitated in young people. A brief overview of these strategies is provided later.

Talking With Young People About Their Feelings

The power of having a trusted person to be able to talk with and share climate-related concerns cannot be overstated (Sanson et al., 2022). Adults, including both parents and teachers, play an important role in supporting young people with climate-related worries and concerns, including feelings that may be causing distress, as well as ways to cope and engage with the issue (Sanson et al., 2022). However, available research suggests that we have some way to go in adequately supporting conversations about climate change. For example, the large multi-country study by Hickman et al. (2021) found that 48% of young people who talked with others about climate change felt that they were ignored or dismissed. It is also imperative that such conversations include not only emotions such as worry and anxiety but also other emotions such as anger, and even positive emotions that may help to facilitate constructive responses to climate change. Such conversations provide an opportunity for adults to share their own feelings on the issue and model adaptive coping skills (Sanson et al., 2022).

Fostering Meaning-Focused Coping and Hope

Sanson et al. (2022) point to the downsides of particular coping strategies such as distancing, in other words trying to ignore the issue of climate change. Strategies involving distancing and avoidance may provide short-term relief in dealing with distress but do not provide active skills to engage with the problem. The large body of research by Ojala demonstrates the promise of using active coping strategies to tackle climate-related concerns, as discussed further later.

Meaning-focused coping entails the capacity to engage in positive reappraisal and to consider more positive aspects of the fight against climate change alongside the more negative prospects (see Ojala, 2022). For example, this may entail considering factors such as progress being made on climate change action and activism or trust in societal actors to make change. This kind of meaning-focused coping has also been associated with generating more positive feelings such as hope (see Ojala, 2012b); however, that same study showed hope to sometimes be associated with de-emphasizing the seriousness of climate change (more akin to emotion-focused coping). Thus, a distinction has been made between "constructive" hope and "hope based in denial," with the former shown to be predicted by feelings and self-efficacy, and related to environmental engagement (Ojala, 2012b, 2015). Constructive engagement with climate change may be driven by framing the problem of climate change in terms of reasons to believe in progress being made towards climate mitigation and promoting feelings of self-efficacy, while still acknowledging the seriousness of the threat. However, it is important to note that positive feelings alone may be indicative of false hope or denial.

Research is needed to examine the effectiveness of meaning-focused programmes on helping young people to manage distressing feelings associated with climate change, as well as how such approaches lead to pro-environmental activities. One study examined outcomes following an intervention where young people participated in a "Youth Climathon," where teams worked together on climate problems (Sarrasin et al., 2022). Four Climathons were examined including 169 young people aged 12 to 18. Key findings were that both self- and collective efficacy, as well as climate anxiety, were associated with higher pro-environmental intentions. A limitation of this work is that questionnaires were only completed by youth post-Climathon; therefore, it is unknown how effective the programme was in helping young people to regulate anxiety and increase pro-environmental intentions. Furthermore, potentially important emotions such as anger were not examined.

Other Strategies

There are a range of other actions that may facilitate coping and engagement in climate-related issues. Parents and teachers can play an important role by modelling pro-environmental action and attitudes. For example, a 12-year longitudinal study found that young people with mothers with greater pro-environmental attitudes engaged in higher levels of pro-environmental behaviours in early adulthood (Evans et al., 2018). This study also found that the time spent outside in childhood was associated with greater environmentally responsible behaviour in adulthood (Evans et al., 2018). This is consistent with other studies that have found that contact or connection with nature has been identified as a factor which can increase pro-environmental attitudes in young people (see Collado & Evans, 2019; Chawla & Gould, 2020). Intriguingly, Collado and Evans (2019) found "outcome expectancy" to be a moderator of the relationship between contact with nature and environmental behaviour. Outcome expectancy shares conceptual similarities with the notion of efficacy (noted earlier as a component of meaning-focused coping). In particular, it appears very similar to the notion of response efficacy, previously examined in the context of pro-environmental behaviour. Thus, there appear to be some overlaps or convergence in the kinds of factors that are relevant to coping with climate change, suggesting some potential for aligning those different perspectives.

Furthermore, engaging in collective action activities such as school strikes and child-led litigation could increase stress and anxiety as noted earlier in the chapter (Ediz & Yanik, 2023). Perhaps engaging in the collective action activities while simultaneously engaging in meaning-focus coping strategies may be a way of somewhat mitigating associated stress. However, as noted earlier, intervention research is sorely needed in this area. There is also huge potential in school-based education programmes in facilitating constructive engagement with climate change.

Future Research Directions

Much of the research on climate worry to date has been conducted with general population samples, often within relatively wealthy Western countries (see Ray, 2021). The kind of climate worry assessed in these studies is typically that which is driven by media and educational exposure to the threat of climate change, which is appropriate given that for most people, exposure to climate change is indirect. We know, however, that many people around the globe are increasingly being exposed to the direct impacts of climate change via various forms of extreme weather and natural disasters. There does seem to be evidence those who are exposed to the direct impacts of climate change (e.g. extreme weather events) do suffer mental health consequences (see Burke et al., 2018; Treble et al., 2023 for a review). Thus, responses to young people who are concerned with climate change will likely need to be catered to the level of direct exposure to the effects of climate change.

As noted by Treble et al. (2023), many studies conducted with "young people" have included adolescents and young adults, rather than younger children. For example, the multi-country study by Hickman et al. (2021) – often cited as an overall indication of rates of climate worry among young people – was conducted with 16–25-year-old participants. Thus, younger children are underrepresented in the data on climate worry (see also Sanson et al., 2019). While some of the findings described earlier may apply equally well to younger age groups, we are not able to reach firm conclusions about this due to a lack of research evidence.

A further consideration is the likelihood of a range of responses to climate change from young people, necessitating a suite of possible responses. Two notable studies have used person-centred techniques to understand young people's responses to climate change. As reviewed earlier, Sciberras and Fernando (2022) examined trajectories of climate worry over time and identified several different patterns, including increasing and decreasing levels of concern. Furthermore, Veijonaho et al. (2023) examined various patterns of reaction to climate change and found four profiles which differed in their levels of climate distress and denialism, which in turn predicted differences in their engagement in pro-environmental behaviour.

Conclusion

Observing the effects of a changing climate all around us will undoubtedly elicit a range of negative feelings, especially for young people thinking about what their futures may hold. Yet, the evidence we have suggested that while worry about climate change can certainly have negative mental health effects, it can also be turned towards more constructive engagement with the climate crisis and effective coping with those negative feelings. Through better understanding the ways in which climate change worry is experienced, and the factors which may facilitate better coping, parents and teachers can support young people to harness their emotions constructively to engage with climate-related issues.

Note

1 Countries included in the study were Australia, Brazil, Finland, France, India, Nigeria, the Philippines, Portugal, the UK, and the USA.

References

Boluda-Verdu, I., Senent-Valero, M., Casas-Escolano, M., Matijasevich, A., & Pastor-Valero, M. (2022). Fear for the future: Eco-anxiety and health implications, a systematic review. *Journal of Environmental Psychology, 84*. https://doi.org/10.1016/j.jenvp.2022.101904

Bouman, T., Verschoor, M., Albers, C. J., Böhm, G., Fisher, S. D., Poortinga, W., & Steg, L. (2020). When worry about climate change leads to climate action: How values, worry and personal responsibility relate to various climate actions. *Global Environmental Change, 62*. https://doi.org/10.1016/j.gloenvcha.2020.102061

Brophy, H., Olson, J., & Paul, P. (2023). Eco-anxiety in youth: An integrative literature review. *International Journal of Mental Health Nursing, 32*(3), 633–661.

Burke, S. E., Sanson, A. V., & Van Hoorn, J. (2018). The psychological effects of climate change on children. *Current Psychiatry Reports, 20*(5), 1–8. https://doi.org/10.1007/s11920-018-0896-9

Chawla, L., & Gould, R. (2020). Childhood nature connection and constructive hope: A review of research on connecting with nature and coping with environmental loss. *People and Nature, 2*(3), 619–642. https://doi.org/10.1002/pan3.10128

Clayton, S. (2020). Climate anxiety: Psychological responses to climate change. *Journal of Anxiety Disorders, 74*, 102263. https://doi.org/10.1016/j.janxdis.2020.102263

Clayton, S., & Karazsia, B. T. (2020). Development and validation of a measure of climate change anxiety. *Journal of Environmental Psychology, 69*. https://doi.org/10.1016/j.jenvp.2020.101434

Collado, S., & Evans, G. W. (2019). Outcome expectancy: A key factor to understanding childhood exposure to nature and children's pro-environmental behavior. *Journal of Environmental Psychology, 61*, 30–36.

Curll, S. L., Stanley, S. K., Brown, P. M., & O'Brien, L. V. (2022). Nature connectedness in the climate change context: Implications for climate action and mental health. *Translational Issues in Psychological Science, 8*(4), 448–460. https://doi.org/10.1037/tps0000329

Ediz, Ç., & Yanik, D. (2023). The effects of climate change awareness on mental health: Comparison of climate anxiety and hopelessness levels in Turkish youth. *International Journal of Social Psychiatry, 69*(8), 2157–2166. https://doi.org/10.1177/00207640231206060

Evans, G. W., Otto, S., & Kaiser, F. (2018). Childhood origins of young adult environmental behaviour. *Psychological Science, 29*, 679–687. https://doi.org/10.1177/095679761774

Frijda, N. H. (1993). Moods, emotion episodes, and emotions. In M. Lewis & J Haviland (Eds.), *Handbook of emotions* (pp. 381–403). Guilford Press.

Gago, T., & Sá, I. (2021). Environmental worry and wellbeing in young adult university students. *Current Research in Environmental Sustainability, 3*, 100064. https://doi.org/10.1016/j.crsust.2021.100064

Gago, T., Sargisson, R. J., & Milfont, T. L. (2024). A meta-analysis on the relationship between climate anxiety and wellbeing. *Journal of Environmental Psychology, 94*, 102230. https://doi.org/10.1016/j.jenvp.2024.102230

Galway, L. P., & Field, E. (2023). Climate emotions and anxiety among young people in Canada: A national survey and call to action. *The Journal of Climate Change and Health, 9*, 100204. https://doi.org/10.1016.j.joclim.2023.100204

Gregersen, T., Doran, R., Böhm, G., & Poortinga, W. (2021). Outcome expectancies moderate the association between worry about climate change and personal energy saving behaviors. *PLoS ONE, 16*(5). https://doi.org/10.1371/journal.pone.0252105

Harth, N. S. (2021). Affect, (group-based) emotions, and climate change action. *Current Opinion in Psychology, 42*, 140–144. https://doi.org/10.1016/j.copsyc.2021.07.018

Hickman, C., Marks, E., Pihkala, P., Clayton, S., Lewandowski, E. R., Mayall, E. E., Wray, B., Mellor, C., & van Susteren, L. (2021). Young people's voices on climate anxiety, government betrayal and moral injury: A global phenomenon. *Lancet Planetary Health, 5*(12), e863–e873. https://doi.org/10.1016/S2542-5196(21)00278-3

Leiserowitz, A., Carman, J., Buttermore, N., Wang, X., Rosenthal, S., Marlon, J. R., & Mulcahy, K. (2021). *International public opinion on climate change.* Yale Program on Climate Change Communication. https://climatecommunication.yale.edu/publications/international-public-opinion-on-climate-change/4/

McBride, S. E., Hammond, M. D., Sibley, C. G., & Milfont, T. L. (2021). Longitudinal relations between climate change concern and psychological wellbeing. *Journal of Environmental Psychology, 78.* https://doi.org/10.1016/j.jenvp.2021.101713

McNeill, I. M., & Dunlop, P. D. (2016). Development and preliminary validation of the constructive and unconstructive worry questionnaire: A measure of individual differences in constructive versus unconstructive worry. *Psychological Assessment, 28*(11), 1368–1378. https://doi.org/10.1037/pas0000266

Ogunbode, C. A., Doran, R., Hanss, D., Ojala, M., Salmela-Aro, K., van den Broek, K. L., Bhullar, N., Aquino, S. D., Marot, T., Schermer, J. A., Wlodarczyk, A., Lu, S., Jianh, F., Maran, D. A., Yadav, R., Ardi, R., Chegeni, R., Ghanbarian, E., Zand, S., . . . Karasu, M. (2022). Climate anxiety, wellbeing and pro-environmental action: Correlates of negative emotional responses to climate change in 32 countries. *Journal of Environmental Psychology, 84.* https://doi.org/10.1016/j.jenvp.2022.101887

Ogunbode, C. A., Pallesen, S., Böhm, G., Doran, R., Bhullar, N., Aquino, S., Marot, T., Schermer, J. A., Wlodarczyk, A., Lu, S., Jiang, F., Salmela-Aro, K., Hanss, D., Maran, D. A., Ardi, R., Chegeni, R., Tahir, H., Ghanbarian, E., Park, J., . . . Lomas, M. J. (2021). Negative emotions about climate change are related to insomnia symptoms and mental health: Cross-sectional evidence from 25 countries. *Current Psychology, 42,* 845–854. https://doi.org/10.1007/s12144-021-01385-4

Ojala, M. (2007). Confronting macrosocial worries: Worry about environmental problems and proactive coping among a group of volunteers. *Futures, 39,* 729–745. https://doi.org/10.1016/j.futures.2006.11.007

Ojala, M. (2012a). How do children cope with global climate change? Coping strategies, engagement, and well-being. *Journal of Environmental Psychology, 32*(3), 225–233. https://doi.org/10.1016/j.jenvp.2012.02.004

Ojala, M. (2012b). Regulating worry, promoting hope: How do children, adolescents, and young adults cope with climate change? *International Journal of Environmental and Science Education, 7*(4), 537–561.

Ojala, M. (2013). Coping with climate change among adolescents: Implications for subjective well-being and environmental engagement. *Sustainability, 5*(5), 2191–2209. https://doi.org/10.3390/su5052191

Ojala, M. (2015). Hope in the face of climate change: Associations with environmental engagement and student perceptions of teachers' emotion communication style and future orientation. *The Journal of Environmental Education, 46*(3), 133–148. https://doi.org/10.1080/00958964.2015.1021662

Ojala, M. (2022). Commentary: Climate change worry among adolescents – on the importance of going beyond the constructive–unconstructive dichotomy to explore coping efforts – a commentary on Sciberras and Fernando (2021). *Child and Adolescent Mental Health, 27*(1), 89–91. https://doi.org/10.1111/camh.12530

Ojala, M. (2023). How do children, adolescents, and young adults relate to climate change? Implications for developmental psychology. *European Journal of Developmental Psychology, 20*(6), 929–943. https://doi.org/10.1080/17405629.2022.2108396

Pihkala, P. (2020). Anxiety and the ecological crisis: An analysis of eco-anxiety and climate anxiety. *Sustainability, 12*(19), 7836.

Ramadan, R., Randell, A., Lavoie, S., Gao, C. X., Manrique, P. C., Anderson, R., McDowell, C., & Zbukvic, I. (2023). Empirical evidence for climate concerns, negative emotions and climate-related mental ill-health in young people: A scoping review. *Early Intervention in Psychiatry, 13374.* https://doi.org/10.1111/eip.13374.

Ray, S. J. (2021). Climate anxiety is an overwhelmingly white phenomenon. *Scientific American.* www.scientificamerican.com/article/the-unbearable-whiteness-of-climate-anxiety/

Reyes, M. E. S., Carmen, B. P. B., Luminarias, M. E. P., Mangulabnan, S. A. N. B., & Ogunbode, C. A. (2021). An investigation into the relationship between climate change anxiety and mental health among Gen Z Filipinos. *Current Psychology, 42*(9), 1–9. https://doi.org/10.1007/s12144-021-02099-3

Sangervo, J., Jylhä, K. M., & Pihkala, P. (2022). Climate anxiety: Conceptual considerations, and connections with climate hope and action. *Global Environmental Change, 76,* 102569. https://doi.org/10.1016/j.gloenvcha.2022.102569

Sanson, A. V., Padilla Malca, K. V., Van Hoorn, J. L., & Burke, S. E. L. (2022). *Children and climate change.* Cambridge University Press.

Sanson, A. V., Van Hoorn, J., & Burke, S. E. (2019). Responding to the impacts of the climate crisis on children and youth. *Child Development Perspectives, 13*(4), 201–207.

Sarrasin, O., Henry, J. L. A., Masserey, C., & Graff, F. (2022). The relationships between adolescents' climate anxiety, efficacy beliefs, group dynamics, and pro-environmental behavioral intentions after a group-based environmental education intervention. *Youth, 2*(3), 422–440. https://doi.org/10.3390/youth2030031

Scherer, K. R. (1984). On the nature and function of emotion: A component process approach. In K. R. Schere & P. Ekman (Eds.), *Approaches to emotion* (pp. 317–331). Psychology Press.

Schmitt, M. T., Aknin, L. B., Axsen, J., & Shwom, R. L. (2018). Unpacking the relationships between pro-environmental behavior, life satisfaction, and perceived ecological threat. *Ecological Economics, 143,* 130–140. https://doi.org/10.1016/j.ecolecon.2017.07.007

Schwartz, S. E. O., Benoit, L., Clayton, S., Parnes, M. K. F., Swenson, L., & Lowe, S. R. (2022). Climate change anxiety and mental health: Environmental activism as a buffer. *Current Psychology,* 1–14. https://doi.org/10.1007/s12144-022-02735-6

Sciberras, E., & Fernando, J. W. (2022). Climate change-related worry among Australian adolescents: An eight-year longitudinal study. *Child and Adolescent Mental Health, 27*(1), 22–29. https://doi.org/10.1111/camh.12521

Smith, N., & Leiserowitz, A. (2014). The role of emotion in global warming policy support and opposition. *Risk Analysis, 34*(5), 937–948. https://doi.org/10.1111/risa.12140

Stanley, S. K., Hogg, T. L., Leviston, Z., & Walker, I. (2021). From anger to action: Differential impacts of eco-anxiety, eco-depression, and eco-anger on climate action and wellbeing. *The Journal of Climate Change and Health, 1.* https://doi.org/10.1016/j.joclim.2021.100003

Swim, J. K., Aviste, R., Lengieza, M. L., & Fasano, C. J. (2022). OK Boomer: A decade of generational differences in feelings about climate change. *Global Environmental Change, 73,* 102479. https://doi.org/10.1016/j.gloenvcha.2022.102479

Thomaes, S., Grapsas, S., van de Wetering, J., Spitzer, J., & Poorthuis, A. (2023). Green teens: Understanding and promoting adolescents' sustainable engagement. *One Earth, 6*(4), 352–361. https://doi.org/10.1016/j.oneear.2023.02.006

Treble, M., Cosma, A., & Martin, G. (2023). Child and adolescent psychological reactions to climate change: A narrative review through an existential lens. *Current Psychiatry Reports,* 1–7. https://doi.org/10.1007/s11920-023-01430-y

Veijonaho, S., Ojala, M., Hietajarvi, L., & Salmela-Aro, K. (2023). Profiles of climate change distress and climate denialism during adolescence: A two-cohort longitudinal study. *International Journal of Behavioural Development, 48*(2), 103–112. https://doi.org/10.1177/01650254231205251.

Vercammen, A., Oswald, T., & Lawrance, E. (2023). Psycho-social factors associated with climate distress, hope and behavioural intentions in young UK residents. *PLoS Global Public Health, 3*(8), e0001938. https://doi.org/10.1371/journal.pgph.0001938

Verlie, B., Clark, E., Jarrett, T., & Supriyono, E. (2021). Educators' experiences and strategies for responding to ecological distress. *Australian Journal of Environmental Education, 37*, 132–146. Https://doi.org/10.1017/aee.2020.34

Verplanken, B., Marks, E., & Dobromir, A. I. (2020). On the nature of eco-anxiety: How constructive or unconstructive is habitual worry about global warming? *Journal of Environmental Psychology, 72*, Article 101528. https://doi.org/10.1016/j.jenvp.2020.101528

Wang, S., Leviston, Z., Hurlstone, M., Lawrence, C., & Walker, I. (2018). Emotions predict policy support: Why it matters how people feel about climate change. *Global Environmental Change, 50*, 25–40. https://doi.org/10.1016/j.gloenvcha.2018.03.002

Whitmarsh, L., Player, L., Jiongco, A., James, M., Williams, M., Marks, E., & Kennedy-Williams, P. (2022). Climate anxiety: What predicts it and how is it related to climate action? *Journal of Environmental Psychology, 83*, 101866. https://doi.org/10.1016/j.jenvp.2022.101866

Wullenkord, M. C., & Ojala, M. (2023). Climate-change worry among two cohorts of late adolescents: Exploring macro and micro worries, coping, and relations to climate engagement, pessimism, and well-being. *Journal of Environmental Psychology, 90*, 102093. https://doi.org/10.1016/j.jenvp.2023.102093

4 How Dare You? How Dare I? The Moral Dimension of Climate Change

Vanessa Kulcar and Barbara Juen

Statement of Positionality

The authors are European psychologists and researchers. Therefore, their perspective is shaped by living in the culture of Western, industrialized, and wealthy nations of the Global North. We expect the moral dimension of climate change to be relevant globally, but to centre around different experiences depending on geographical, historical, and cultural backgrounds. Specifically, we assume distress caused by moral transgressions of others to be the primary reaction in the Global South. Within the societies of the Global North, we assume distress caused by moral transgressions by oneself to be prevalent in addition as these societies contributed and continue to contribute significantly to climate change. However, transgressions by others can be experienced globally and particularly among youth. A question remaining unanswered this far is how these perceived transgressions by oneself and by others interact and how resulting mental health strains differ between different regions of the world.

The Moral Dimension of Climate Change

> *You have stolen my dreams and my childhood with your empty words. And yet I'm one of the lucky ones. People are suffering. People are dying. Entire ecosystems are collapsing. We are in the beginning of a mass extinction. And all you can talk about is money and fairy tales of eternal economic growth. How dare you!*
>
> (Greta Thunberg, 2019, September 23)

Greta Thunberg clearly pointed out the moral element of climate-relevant behaviour during her speech in front of global decision makers at the 2019 Climate Action Summit. She specifically addressed people in power and described their failure to mitigate climate change as morally wrong. With this point of view, she is not alone. A recent survey by Hickman and colleagues (2021) provided evidence of a perception of betrayal by political decision makers among young

DOI: 10.4324/9781003494416-5

people worldwide: A majority perceived people in power as failing and betraying young people and future generations while dismissing the distress youth are experiencing. In addition to decisions by people in power, the behaviour of all members of societies, including each individual's own behaviour, is relevant for climate change and therefore gets morally evaluated (Fløttum et al., 2016; Henritze et al., 2023). This moral evaluation impacts a range of outcomes, including climate-relevant behaviour, mental well-being, and perspectives on fairness, humanity, and the future (Bonson et al., 2023; Doran et al., 2019; Feinberg & Willer, 2011; Hickman et al., 2021; Pearson et al., 2021). Therefore, it is crucial to consider morality in both practice and research related to climate change and other sustainability issues.

Climate change is affecting the lives and livelihoods of people and non-human beings around the world and will impact younger and future generations increasingly. These impacts are based on different mechanisms, including more frequent and intense extreme weather events, and disturbances of economic, social, and political systems with consequences for the humans living within these systems (Jamieson, 2007). Thereby, climate change threatens fundamental human rights, including the right to life, health, and subsistence (Caney, 2010). Non-human lives and nature in itself are also harmed by climate change (Gardiner, 2006; Jamieson, 2007). Accepting this preventable harm to other beings can be considered a failure to adhere to basic moral principles such as "do no harm" (Gardiner, 2006). Therefore, climate change is a morally relevant topic (Caney, 2010; Gardiner, 2006; Singer, 2006). Before anthropogenic climate change was understood, behaviours contributing to greenhouse gas emissions were considered as morally neutral (Bostrom et al., 2020; Li et al., 2021). Once the consequences of such behaviours become known, they become subject to moral evaluation.

The moral dimension of climate change highlights global, social, and intergenerational injustice (Antadze, 2020; Asayama et al., 2021; Barnwell & Wood, 2022; Henritze et al., 2023). Climate change is unjust as countries and generations contributed differently to its causation, and the greatest burden of climate change falls on those who have contributed least (Barnwell & Wood, 2022; Ingle & Mikulewiicz, 2020). This is partly based on geographical differences but also on historically grown inequality and associated differences in the ability of individuals and communities to respond to climate change hazards (Barnwell & Wood, 2022). Thus, climate change not only builds upon inequality but further perpetuates global disparities (Asayama et al., 2021; Henritze et al., 2023).

However, considering climate change from a moral perspective comes with complications. The moral relevance of climate change easily gets dismissed due to characteristics such as the unobtrusive relations between causes and effects or unintended causation (Gardiner, 2006; Markowitz & Shariff, 2012). Additionally, individual and collective interests diverge as individuals (including individual industries, nations, etc.) profit from not limiting their emissions, even though climate change mitigation is the collective imperative (Gardiner, 2006; Singer, 2006). Climate change, therefore, is only manageable if humans decide to adhere to moral principles and collectively put efforts into climate change mitigation (Dunn, 2021; Meijers et al., 2021). For this reason, there are approaches to highlight the moral dimension of climate change to support these efforts, for example, by making climate change impacts tangible in a morally relevant way (Markowitz & Shariff, 2012; Nolt, 2015).

How Do People Come to Perceive Climate Change as Morally Relevant?

To perceive climate-relevant behaviour as a moral issue, it has to be moralized. Moralization describes the process through which something previously neutral becomes perceived as morally relevant (Feinberg et al., 2019). This can happen at the level of individuals and also at the levels of groups or whole societies (Rhee et al., 2019; Rozin, 1999). Moralization can be promoted by associating climate change – and climate-relevant behaviour – with damage and harm on a cognitive level (Feinberg et al., 2019; Skitka et al., 2021). For example, people might realize that climate change increases the frequency and severity of disasters and that their mobility behaviour perpetuates climate change by emitting carbon dioxide (CO_2). This realization can be further supported by the experience of moral emotions (Feinberg et al., 2019; Skitka et al., 2021), for example, feeling guilty when taking the car but proud when taking public transportation. However, moralization can also be hindered (Feinberg et al., 2019). For example, people might perceive public transportation as tedious, or the social norm is to take the car. This can contribute to efforts to justify and rationalize car usage or even the deflection of all messages that highlight the threat of climate change and the damage caused by anthropogenic greenhouse gas emissions. Responsibility for climate change and thereby moral relevance of climate-relevant behaviour is further deflected by focusing on the role of the collective and systems, thereby minimizing individual impact (Caillaud et al., 2016). As these mechanisms act differently among individuals, there can be large differences in the moralization of different climate-relevant behaviours among the population.

Once a behaviour is moralized, it is perceived as either fundamentally right or wrong (Skitka et al., 2021). For example, if a person perceives climate activism as morally relevant behaviour, they will feel internally motivated and obligated to participate in activist behaviours. Usually, these convictions also extend to other people, and individuals with strong moral convictions are unwilling to accept deviant beliefs or behaviours (Skitka, 2010). As climate change disproportionately affects those who contributed least to its causation (Barnwell & Wood, 2022), and wrongfully inflicting harm on innocent and marginalized groups is usually evaluated as morally wrong (Rozin, 1999), the moral relevance of climate change-related behaviours is further increased.

Accordingly, a study among four European countries in 2016 indicated that the majority of the population expressed some degree of moral concern regarding climate change (Doran et al., 2019). Similarly, 83% of young Norwegians perceived a moral obligation to mitigate climate change in rich nations, and 59% perceived a moral obligation to do so as individuals (Fløttum et al., 2016). Accordingly, reasoning for climate change mitigation and adaptation is often based on moral arguments (Adger et al., 2017).

Consequences of Moralization

If behaviours or entities are moralized, they are fundamentally treated differently. People are committed more strongly to a cause based on moral convictions compared to pure preference (Skitka et al., 2021). In the context of climate change, moral concern is associated with increased support for climate change mitigation (Doran et al., 2019). There is a significant body of evidence indicating increased efforts towards climate change mitigation and adaptation, as well as environmental protection, based on moral emotions such as compassion (Lu & Schuldt, 2016), disgust (Jia et al., 2017), pride and guilt (Onwezen et al., 2013; Schneider et al., 2017), or based on moral identities (Jia et al., 2017).

The association between moralization and behaviour is also reflected on a broader societal level (Rozin, 1999): Governments are more likely to pass policies in line with societal moral convictions; institutions like schools promote them further; and moral convictions are passed on between people and generations. Consequently, procedural aspects of decision making become less relevant, and political outcomes are evaluated purely based on adherence to moral standards (Ryan, 2019). As people seek acceptance and positive evaluation, they try to adhere to their group's standards (Haidt, 2001), so moralized behaviours are enforced both externally and internally.

However, moral evaluations can be context dependent as shown in a study on Chinese adolescents (Sternäng & Lundholm, 2011). In group interviews, the young people held unknown others to higher standards than themselves concerning climate change mitigation. For themselves, they focused on personal interests, dismissing potential environmental impacts or highlighting technological developments as pathways to a more sustainable future. Contrarily, climate change perpetuating behaviours of others were judged to be the result of low moral standards which lead to calls for penalties.

Moral Transgressions in the Context of Climate Change

Moral transgressions are characterized by violations of perceived moral standards. In the context of climate change, they can be differentiated by their focus on the behaviour of others versus one's behaviour (Henritze et al., 2023). Moral transgressions by others largely focus on people in positions of power – for example, political decision makers, or people in the Global North – perpetuating climate change and failing to establish mitigation strategies. The high prevalence of perceived moral transgression by others among youth has been established by recent research (Hickman et al., 2021): Young people reported feeling betrayed by governments and a majority perceived decision makers as lying and failing young people and future generations. This feeling of betrayal is associated with negative emotions (Hickman et al., 2021). The perception that people in power neglect environmental protection is not a new development (e.g. Fløttum et al., 2016; Leahy et al., 2010) but might be exacerbated as climate change advances. Additionally, not only people in positions of political power but humans in general are often perceived as unwilling to act morally correctly in the context of environmental issues (Fischer et al., 2011).

Climate-conscious individuals do not always act in accordance with their ideals either (Henritze et al., 2023; Weintrobe, 2020). Especially in the Global North, people are living in a society that encourages climate-damaging behaviours. Their culture reinforces people behaving in ways that harm the environment, other people, and living beings by suppressing awareness of and reflection on the consequences of their actions (Weintrobe, 2020). Weintrobe (2020) argues that we are living in "bubbles" that depict harming behaviours as the only alternative and shield us from awareness of negative consequences associated with the harming behaviour. However, these bubbles can burst, and becoming aware of one's own moral transgressions can be challenging. Further, wanting to live sustainably is attached to ambiguities, for example, the individual can be motivated to protect the environment but simultaneously feel that they will not be able to make a difference on their own (Ojala & Anniko, 2020). This can be perpetuated by focusing on individual instead of collective responsibility for climate change (Henritze et al., 2023). As young people have a reduced scope of action against climate change (Martin et al., 2022), they might be particularly vulnerable to perceiving moral transgressions within their own behaviour.

Mental Health Implications of Moral Transgressions

Recently, researchers started to consider moral transgressions in the context of climate change as resulting in moral injury (Henritze et al., 2023; Hickman et al., 2021). Moral injury was originally researched in war veterans and was described as the result of witnessing or committing atrocities (Jinkerson, 2016; Litz & Kerig, 2019; Shay, 2014). Later, the concept was expanded to other populations, such as healthcare workers, refugees, or child protection services (Griffin et al., 2019; Ter Heide & Olff, 2023). The mental health implications of moral injury are characterized by a disruption of fundamental beliefs, including the belief in the good in oneself, people, and the world (Bonson et al., 2023; Litz et al., 2009). Moral injury is associated with guilt and shame, as well as a loss of trust in oneself and others (Jinkerson, 2016), and has negative impacts on emotional, psychological, behavioural, spiritual, and social dimensions (Litz et al., 2009). People affected tend to withdraw from others and neglect to care for themselves, which further perpetuates negative beliefs and expectations. Thereby, moral injury has been demonstrated to be associated with poor mental health outcomes (Griffin et al., 2019; Jinkerson, 2016).

Litz and Kerig (2019) differentiated moral injury from the impacts of other events that are perceived as morally wrong. They proposed a continuum model of mental health impacts resulting from moral transgressions based on their prevalence within the population, the severity of associated impairment, and the frequency of eliciting events. Frequent events were defined as *moral challenges* and assumed to cause only mild mental strains that were termed *moral frustration*. According to Litz and Kerig (2019), climate change represents such a moral challenge, potentially causing moral frustration. Moral frustration is not assumed to have severe mental health impacts. Moral injury, on the other hand, is described as a severe impairment based on an extreme event, with overall low prevalence in a population. As the frequency of triggering events and the prevalence within the population are described as low, climate change cannot cause moral injury based on Litz's and Kerig's definition since it affects a wide range of people globally.

Alternative concepts that can be employed to understand the impacts of moral transgressions on mental health are moral distress and associated moral residue. These concepts were developed in the context of healthcare settings (Epstein & Delgado, 2010). Healthcare workers experience moral distress when they are forced to act against their moral values (Epstein & Delgado, 2010; Ten Have & Patrão Neves, 2021a). Moral distress is an acute reaction in itself. However, it can linger in the form of moral residue (Epstein & Delgado, 2010; Ten Have & Patrão Neves, 2021b). Similar to moral injury, these reactions are associated with impairments of emotional well-being (Rodney, 2017). If there are repeated moral transgressions, and moral concerns are not appropriately considered by the organization or superiors, moral residue can accumulate and threaten a person's moral integrity and self-perception. Moral residue, therefore, has parallels to moral injury.

The concepts of moral injury and moral distress were originally based on research on veterans or healthcare workers who are exposed to life-or-death situations. In comparison, transgressions related to climate change, including, for example, air travel or a lack of climate protection policies, appear less profound and disruptive. However, it is argued that there might be a cumulative effect of frequent everyday transgressions (Brenner, 2017). A similar so-called crescendo effect is assumed in the case of moral distress and moral residue (Epstein & Delgado, 2010). The regular moral transgressions regarding climate change can thereby be expected to have similar impacts as singular but extreme transgressions. Additionally, research on moral injury and moral distress to date has primarily focused on adults in their work environments. Children and adolescents who are still developing might react more sensitively to mental health impacts of climate change (Vergunst & Berry, 2022), in particular, since youth is the time during which most mental health problems arise (Solmi et al., 2022). Indeed, they might react more strongly to everyday transgressions related to climate change, which could lead to mental health impacts to a similar extent as moral injury in veterans or moral distress in healthcare workers. Researchers argue that the defining factor of moral injury is not solely the eliciting event but rather the impacts that individuals experience (Bonson et al., 2023). Therefore, moral injury and moral distress are potential outcomes of such everyday transgressions, and demands to broaden the concepts to more than isolated extreme events have emerged (Brenner, 2017; Dunn, 2021).

What Is Needed in Research?

Different versions of moral distress and moral injury are increasingly addressed in research and in clinical psychotraumatology with research growing exponentially during the last decade (Griffin et al., 2019; Ter Heide & Olff, 2023). This growing body of research denotes the significance of moral suffering but also moral well-being within mental health (Rodney, 2017; Ter Heide & Olff, 2023). However, clear definitions and conceptualizations of moral injury, moral distress, and other reactions to perceived moral transgressions are still lacking (Griffin et al., 2019; Rodney, 2017; Ter Heide & Olff, 2023). Advancing definitions and conceptualizations is necessary to further research in the context of climate change.

Overall, clear definitions and delimitations between different emotional reactions in response to climate change are pending. Thus, it is necessary to elaborate on how emotional reactions interact and whether they should be conceptualized as distinct or as subtypes of larger constructs. For instance, moral emotions could be viewed as one aspect of a broader category such as climate anxiety (see Hickman et al., 2021) or climate change distress. In that sense, moral emotions like guilt, shame, or outrage in response to behaviour that contributes to climate change would be part of a broader response to climate change, which also includes emotions such as anxiety, fear, or grief. There could also be an overarching construct of climate change distress which may include different subdimensions, with moral emotions being one of those. Alternatively, moral distress or injury in response to climate change could be conceptualized separately from other reactions to climate change. At this point, more theoretical considerations and empirical research are necessary to enhance our understanding of various aspects of psychological responses, their interrelationships, and their definitions and operationalization.

The matter of definitions also involves distinguishing between pathological reactions that require psychotherapeutic interventions and adaptive reactions to witnessed moral problems that may support people in addressing injustice

(Pihkala, 2020; Ter Heide & Olff, 2023). Currently, the boundaries between pathological moral injury as a profound disruption of beliefs and relationships (Bonson et al., 2023; Litz et al., 2009) or overwhelming and paralysing climate anxiety (Pihkala, 2020) and adaptive reactions to the threat and to everyday moral transgressions are blurred. As moral injury is discussed as a potential diagnosis similar to post-traumatic stress disorder or as a subtype (Ter Heide & Olff, 2023), referring to the reaction to moral transgressions in the context of climate change as moral injury may inappropriately pathologize youth.

Research is needed specifically on reactions to moral transgressions in the context of climate change. This research should encompass different dimensions, including transgressions by individuals themselves and by others, and it should consider different populations. This includes people from different regions of the world who are affected differently by climate change, for example, from the Global North and the Global South. Additionally, different age groups need to be studied to take moral development during childhood and adolescence into account. Children as young as three years old perceive behaviour that harms the environment as morally wrong (Hahn & Garret, 2017) and moral evaluations develop to be more nuanced and complex as they grow up. For example, they begin to differentiate between harming the environment and harming people (Hahn & Garret, 2017) and consider additional social factors in their judgements (Nucci & Turiel, 2009). Moral evaluations fluctuate between these developmental periods and convictions regarding the environment tend to weaken in late adolescence (Krettenauer, 2017). Therefore, differences in moral development need to be considered in research on mental health implications of the moral dimension of climate change. However, regardless of their age, youth react emotionally to their own environmentally relevant behaviour (Krettenauer, 2017), highlighting their vulnerability to the mental health impacts of moral transgressions.

So far, research on the moralization of climate change has primarily focused on its impact on climate-relevant behaviour (e.g. Doran et al., 2019; Lu & Schuldt, 2016; Onwezen et al., 2013; Schneider et al., 2017). It is crucial to expand the scope to potential mental health impacts as the crisis progresses and evidence of moral struggles accumulates. Risk factors for mental health impairments should be examined to identify people in need of support. Youth should be a particular focus as they are vulnerable to the mental health impacts of climate change and may suffer more severely from moral distress due to their own future being affected and their limited scope of action (Martin et al., 2022; Ojala, 2015a). Research on mental health impacts of moral transgressions should contribute to interventions that alleviate mental health strains without inhibiting the motivational component of moral convictions. It is necessary to find a balance that enables people to recognize the injustice of climate change and respond to it without feeling overwhelmed by moral violations, and as a result withdraw and feel paralysed.

Within research, intersectional approaches should be chosen. Climate change interacts with historically grown inequities based, for example, on colonialism and racism (Barnwell & Wood, 2022). It is crucial to consider these dimensions in research on the mental health implications of climate change, especially with regard to moral distress, as people from different backgrounds may perceive the moral implications of climate change differently. Barnwell and Wood (2022) argue that psychological distress related to climate change is inherently linked to experiences of discrimination and marginalization based on unjust socio-political systems. These complex interactions need to be addressed in further research to enable adequate support for those most affected by and least equipped to respond to climate change.

What Is Needed From Practitioners Working With Young People?

Mental health impacts of moral transgressions in the context of climate change are not yet well understood, and evidence for effective interventions is lacking. However, measures need be taken today to protect youth from negative impacts and help them handle the strain of exposure. The first step is awareness of potential moral distress in youth by adults working and living with them, such as youth workers, teachers, and parents. Children's experience of climate change emotions is shaped by their social environment (Crandon et al., 2022; Ojala, 2015b) and adults play a crucial role in fostering healthy coping (Sanson et al., 2018; Vamvalis, 2023). If youth experience moral distress, steps to support them need to be taken. Lessons can be learned from addressing potential mental health strains of climate anxiety and from research on therapeutic interventions for moral injury in veterans or moral distress in healthcare workers.

Research on moral injury suggests that contextualizing moral transgressions within external constraints and influences may be helpful in addressing moral transgressions committed by individuals themselves (Griffin et al., 2019). Recognizing that one lives within a society that promotes climate-damaging behaviours can aid in forgiving oneself for not always behaving in climate-friendly ways. It is crucial to provide an environment without judgement in which behaviours are understood and emotions accepted (Griffin et al., 2019; Serfioti et al., 2023). Central concepts to be included when addressing moral distress or moral injury are acceptance and (self-) compassion (Serfioti et al., 2023). Sometimes, imagining a moral authority – for example, a spiritual authority, but in case of climate change possibly a climate activist such as Greta Thunberg – and their compassionate and forgiving reaction to one's transgressions can be helpful (Serfioti et al., 2023).

Such a comprehension of external influences and a focus on compassion may also promote understanding of others, and thereby set a baseline for communication based on empathy instead of blame. This can further be encouraged by teaching perspective-taking with others, for example, via creative outlets such as theatre or art (Ojala, 2016). Additionally, teachers can support their students in regaining trust in other humans by highlighting positive examples and inviting individuals who have contributed to a sustainable world into the classroom (Ojala, 2016; Sanson et al., 2018). Learning about others' efforts to promote positive change can foster hope and mitigate negative perceptions of humanity.

As attributions are decisive for the development of moral injury (Bonson et al., 2023; Litz et al., 2009), addressing maladaptive attributions can be another pathway to mitigate mental health strains (Griffin et al., 2019). These include, for example, overgeneralizations and self-blame (Bonson et al., 2023; Litz et al., 2009). Here, cognitive behavioural approaches can be helpful: for example, reflecting on perceived wrong-doings and counteracting overgeneralization (Griffin et al., 2019). Within the classroom, educators can discuss self-talk about climate-relevant behaviour (Ojala, 2016; Sanson et al., 2018). Adults can support youth identify negative ways they talk to themselves (e.g. blaming themselves for failures to behave in climate-friendly ways) and offer alternative interpretations (e.g. highlighting what they have already achieved). In general, providing space for talking about and reflecting on difficult climate change emotions (Sanson et al., 2018) as well as psycho-education (Vergunst & Berry, 2022) can be helpful for children and adolescents.

Besides general struggles in coping with difficult moral emotions, the experience of moral distress or injury often leads to isolation (Bonson et al., 2023; Serfioti et al., 2023; Shay, 2014). Similar reactions can be expected for moral distress related to climate change. Therefore, it is crucial to maintain or restore social networks and support communication (Epstein & Delgado, 2010; Rodney, 2017; Serfioti et al., 2023). For youth, the classroom is a promising setting to develop social networks and facilitate communication about difficult topics, such as perceived moral failures. Climate change, including climate-relevant behaviour and associated emotions, can be explicitly addressed in a non-judgemental atmosphere that promotes exchange of perceptions and experiences. In addition, connections between particularly involved children and adolescents can be encouraged. This can go beyond the setting of individual classrooms and include, for example, climate clubs that focus on not only facilitating change but also talking about climate change emotions. Connecting with others who have similar experiences can be helpful in overcoming the emotional struggles (Cunsolo et al., 2020; Serfioti et al., 2023; Weintrobe, 2020). Especially youth whose climate concerns are not shared by their immediate surrounding may benefit from connecting with other concerned people. This can help them perceive their emotional reactions as normal and provide a collective experience to cope with mental strain. By working together, they can overcome their struggles and take action against climate change.

Beyond the management of reactions that are elicited in youths' everyday lives, schools play a crucial role in preventing mental struggles by considering the moral and emotional dimensions of climate change education. Climate change education should not exclusively provide knowledge about the scientific mechanisms of climate change but also needs to encompass social and psychological aspects (Vamvalis, 2023). The transfer of knowledge should be combined with opportunities for emotional processing, reflection, and discussion of the information (Trott, 2022). Additionally, educational interventions should be mindful of the message they convey regarding responsibility for action: While highlighting opportunities for action can be beneficial (Sanson et al., 2018), placing the burden of responsibility on youth can be overwhelming (Vamvalis, 2023) and might exacerbate moral struggles. Communication regarding climate change should, therefore, always consider its moral implications and provide a safe space to work through any emotions it may evoke.

To facilitate emotional processing, adults first need to address their own climate emotions (Baker et al., 2021). Only when they are aware of their own emotional state, are they able to provide the necessary space for the emotions of children and adolescents. Otherwise, apparent evasion of the topic might appear to youth as ignoring their concerns and further foster feelings of betrayal and loss of trust (Ojala, 2016). On the other hand, if adults open up about their own emotions, they can serve as role models and normalize talking about and processing one's emotions (Baker et al., 2021).

Overall, addressing the mental health impacts of climate change requires an intersectional approach, taking into account different factors that can lead to inequity and marginalization (Barnwell & Wood, 2022; Pearson et al., 2021). This includes, among others, considering the differential impacts based on gender, race, and social status. Furthermore, as youth from the Global South are disproportionally affected by climate change (Barnwell & Wood, 2022; Vergunst & Berry, 2022), it is crucial to support them specifically. Participatory approaches that include youth in the design and implementation of interventions are promising for addressing their specific needs while simultaneously working within limited resources (Barnwell & Wood, 2022). Examples for participatory actions include involving youth via peer-support methods or providing opportunities to educate their community on topics relevant to them.

In general, moral distress should be addressed on not only an individual level but also the causes for distress – that is, systems fostering climate change – need to be attended to (Levinson, 2015). Negative emotional reactions, including moral distress, can be regarded as a tool to motivate change (Ford & Feinberg, 2020; Marcus, 1980). When moral convictions are violated, people want to protect their values by increasing collective efforts to right perceived wrongs (Pauls et al., 2022). Thus, a balance between harnessing the change-inducing potential of moral emotions and safeguarding against mental health impairments needs to be developed (Levinson, 2015).

Conclusion

On the one hand, negative emotional reactions to moral transgressions can be considered a sign of mental health instead of illness (Weintrobe, 2020). Therefore, harvesting these reactions to motivate action rather than solely focusing on reducing moral distress is foundational for democracies to work and to address injustice and societal problems (Antadze, 2020; Barnwell & Wood, 2022; Ford & Feinberg, 2020). However, youth who have a limited scope of action while being particularly vulnerable to present and future climate change impacts (Vergunst & Berry, 2022) may be overwhelmed by the ongoing failures to address the climate crisis. Research on potential mental health effects of moral transgressions in the context of climate change is scarce to date. Still, findings from other fields predict extensive negative implications for young peoples' mental health, as well as their ability to take action against climate change. Therefore, we urge greater attention to and awareness of negative climate emotions in general and moral emotions in specific. Emotional responses to the topic need to be considered and specifically addressed, both within and beyond climate change education. The classroom should provide a safe space for youth to work through their emotional reactions and moral struggles related to the climate crisis and to receive support from adults and peers.

References

Adger, W. N., Butler, C., & Walker-Springett, K. (2017). Moral reasoning in adaptation to climate change. *Environmental Politics*, *26*(3), 371–390. https://doi.org/10.1080/09 644016.2017.1287624

Antadze, N. (2020). Moral outrage as the emotional response to climate injustice. *Environmental Justice*, *13*(1), 21–26. https://doi.org/10.1089/env.2019.0038

Asayama, S., Emori, S., Sugiyama, M., Kasuga, F., & Watanabe, C. (2021). Are we ignoring a black elephant in the Anthropocene? Climate change and global pandemic as the crisis in health and equality. *Sustainability Science*, *16*(2), 695–701. https://doi.org/10.1007/s11625-020-00879-7

Baker, C., Clayton, S., & Bragg, E. (2021). Educating for resilience: Parent and teacher perceptions of children's emotional needs in response to climate change. *Environmental Education Research*, *5*, 687–705. https://doi.org/10.1080/13504622.2020.1828288

Barnwell, G., & Wood, N. (2022). Climate justice is central to addressing the climate emergency's psychological consequences in the Global South: A narrative review. *South African Journal of Psychology*, *52*(4), 486–497. https://doi.org/10.1177/00812463211073384

Bonson, A., Murphy, D., Aldridge, V., Greenberg, N., & Williamson, V. (2023). Conceptualization of moral injury: A socio-cognitive perspective. *Journal of Military, Veteran and Family Health*, e20220034. https://doi.org/10.3138/jmvfh-2022-0034

Bostrom, A., Böhm, G., Hayes, A. L., & O'Connor, R. E. (2020). Credible threat: Perceptions of pandemic Coronavirus, climate change and the morality and management of global risks. *Frontiers in Psychology*, *11*, 578562. https://doi.org/10.3389/fpsyg.2020.578562

Brenner, G. H. (2017). Considering collective moral injury following the 2016 election. *Contemporary Psychoanalysis*, *53*(4), 547–560. https://doi.org/10.1080/00107530.20 17.1384683

Caillaud, S., Bonnot, V., Ratiu, E., & Krauth-Gruber, S. (2016). How groups cope with collective responsibility for ecological problems: Symbolic coping and collective emotions. *British Journal of Social Psychology, 55*(2), 297–317. https://doi.org/10.1111/bjso.12126

Caney, S. (2010). Climate change, human rights, and moral thresholds. In S. M. Gardiner, S. Caney, D. Jamieson & H. Shue (Eds.), *Climate ethics* (pp. 163–177). Oxford University Press. https://doi.org/10.1093/oso/9780195399622.003.0018

Crandon, T. J., Scott, J. G., Charlson, F. J., & Thomas, H. J. (2022). A social–ecological perspective on climate anxiety in children and adolescents. *Nature Climate Change, 12*(2), 123–131. https://doi.org/10.1038/s41558-021-01251-y

Cunsolo, A., Harper, S. L., Minor, K., Hayes, K., Williams, K. G., & Howard, C. (2020). Ecological grief and anxiety: The start of a healthy response to climate change? *The Lancet Planetary Health, 4*(7), e261–e263. https://doi.org/10.1016/S2542-5196(20)30144-3

Doran, R., Böhm, G., Pfister, H.-R., Steentjes, K., & Pidgeon, N. (2019). Consequence evaluations and moral concerns about climate change: Insights from nationally representative surveys across four European countries. *Journal of Risk Research, 22*(5), 610–626. https://doi.org/10.1080/13669877.2018.1473468

Dunn, S. (2021). Moral injury and tragic sensibility: Reframing the Iraqi refugee crisis. *Journal of Religious Ethics, 49*(3), 462–478. https://doi.org/10.1111/jore.12362

Epstein, E., & Delgado, S. (2010). Understanding and addressing moral distress. *OJIN: The Online Journal of Issues in Nursing, 15*(3). https://doi.org/10.3912/OJIN.Vol15No03Man01

Feinberg, M., Kovacheff, C., Teper, R., & Inbar, Y. (2019). Understanding the process of moralization: How eating meat becomes a moral issue. *Journal of Personality and Social Psychology, 117*(1), 50–72. https://doi.org/10.1037/pspa0000149

Feinberg, M., & Willer, R. (2011). Apocalypse soon?: Dire messages reduce belief in global warming by contradicting just-world beliefs. *Psychological Science, 22*(1), 34–38. https://doi.org/10.1177/0956797610391911

Fischer, A., Peters, V., Vávra, J., Neebe, M., & Megyesi, B. (2011). Energy use, climate change and folk psychology: Does sustainability have a chance? Results from a qualitative study in five European countries. *Global Environmental Change, 21*(3), 1025–1034. https://doi.org/10.1016/j.gloenvcha.2011.04.008

Fløttum, K., Dahl, T., & Rivenes, V. (2016). Young Norwegians and their views on climate change and the future: Findings from a climate concerned and oil-rich nation. *Journal of Youth Studies, 19*(8), 1128–1143. https://doi.org/10.1080/13676261.2016.1145633

Ford, B. Q., & Feinberg, M. (2020). Coping with politics: The benefits and costs of emotion regulation. *Current Opinion in Behavioral Sciences, 34*, 123–128. https://doi.org/10.1016/j.cobeha.2020.02.014

Gardiner, S. M. (2006). A perfect moral storm: Climate change, intergenerational ethics and the problem of moral corruption. *Environmental Values, 15*(3), 397–413. https://doi.org/10.3197/096327106778226293

Griffin, B. J., Purcell, N., Burkman, K., Litz, B. T., Bryan, C. J., Schmitz, M., Villierme, C., Walsh, J., & Maguen, S. (2019). Moral injury: An integrative review. *Journal of Traumatic Stress, 32*(3), 350–362. https://doi.org/10.1002/jts.22362

Hahn, E. R., & Garrett, M. K. (2017). Preschoolers' moral judgments of environmental harm and the influence of perspective taking. *Journal of Environmental Psychology, 53*, 11–19. https://doi.org/10.1016/j.jenvp.2017.05.004

Haidt, J. (2001). The emotional dog and its rational tail: A social intuitionist approach to moral judgment. *Psychological Review, 108*(4), 814–834. https://doi.org/10.1037//0033-295X.108.4.814

Henritze, E., Goldman, S., Simon, S., & Brown, A. D. (2023). Moral injury as an inclusive mental health framework for addressing climate change distress and promoting justice-oriented care. *The Lancet Planetary Health*, *7*(3), e238–e241. https://doi.org/10.1016/S2542-5196(22)00335-7

Hickman, C., Marks, E., Pihkala, P., Clayton, S., Lewandowski, R. E., Mayall, E. E., Wray, B., Mellor, C., & van Susteren, L. (2021). Climate anxiety in children and young people and their beliefs about government responses to climate change: A global survey. *The Lancet Planetary Health*, *5*(12), e863–e873. https://doi.org/10.1016/S2542-5196(21)00278-3

Ingle, H. E., & Mikulewicz, M. (2020). Mental health and climate change: Tackling invisible injustice. *The Lancet Planetary Health*, *4*(4), e128–e130. https://doi.org/10.1016/S2542-5196(20)30081-4

Jamieson, D. (2007). The moral and political challenges of climate change. In S. C. Moser & L. Dilling (Hrsg.), *Creating a climate for change: Communicating climate change and facilitating social change* (pp. 475–482). Cambridge University Press. https://doi.org/10.1017/CBO9780511535871.033

Jia, F., Soucie, K., Alisat, S., Curtin, D., & Pratt, M. (2017). Are environmental issues moral issues? Moral identity in relation to protecting the natural world. *Journal of Environmental Psychology*, *52*, 104–113. https://doi.org/10.1016/j.jenvp.2017.06.004

Jinkerson, J. D. (2016). Defining and assessing moral injury: A syndrome perspective. *Traumatology*, *22*(2), 122–130. https://doi.org/10.1037/trm0000069

Krettenauer, T. (2017). Pro-environmental behavior and adolescent moral development. *Journal of Research on Adolescence*, *27*(3), 581–593. https://doi.org/10.1111/jora.12300

Leahy, T., Bowden, V., & Threadgold, S. (2010). Stumbling towards collapse: Coming to terms with the climate crisis. *Environmental Politics*, *19*(6), 851–868. https://doi.org/10.1080/09644016.2010.518676

Levinson, M. (2015). Moral injury and the ethics of educational injustice. *Harvard Educational Review*, *85*(2), 203–228. https://doi.org/10.17763/0017-8055.85.2.203

Li, Y., Sewell, D. K., Saber, S., Shank, D. B., & Kashima, Y. (2021). The climate commons dilemma: How can humanity solve the commons dilemma for the global climate commons? *Climatic Change*, *164*(1–2), 4. https://doi.org/10.1007/s10584-021-02989-2

Litz, B. T., & Kerig, P. K. (2019). Introduction to the special issue on moral injury: Conceptual challenges, methodological issues, and clinical applications. *Journal of Traumatic Stress*, *32*(3), 341–349. https://doi.org/10.1002/jts.22405

Litz, B. T., Stein, N., Delaney, E., Lebowitz, L., Nash, W. P., Silva, C., & Maguen, S. (2009). Moral injury and moral repair in war veterans: A preliminary model and intervention strategy. *Clinical Psychology Review*, *29*(8), 695–706. https://doi.org/10.1016/j.cpr.2009.07.003

Lu, H., & Schuldt, J. P. (2016). Compassion for climate change victims and support for mitigation policy. *Journal of Environmental Psychology*, *45*, 192–200. https://doi.org/10.1016/j.jenvp.2016.01.007

Marcus, R. B. (1980). Moral dilemmas and consistency. *The Journal of Philosophy*, *77*(3), 121–136.

Markowitz, E. M., & Shariff, A. F. (2012). Climate change and moral judgement. *Nature Climate Change*, *2*(4), 243–247. https://doi.org/10.1038/nclimate1378

Martin, G., Reilly, K., Everitt, H., & Gilliland, J. A. (2022). Review: The impact of climate change awareness on children's mental well-being and negative emotions – a scoping review. *Child and Adolescent Mental Health*, *27*(1), 59–72. https://doi.org/10.1111/camh.12525

Meijers, M. H. C., Scholz, C., Torfadóttir, R. H., Wonneberger, A., & Markov, M. (2021). Learning from the COVID-19 pandemic to combat climate change: Comparing drivers of individual action in global crises. *Journal of Environmental Studies and Sciences,* *12,* 272–282. https://doi.org/10.1007/s13412-021-00727-9

Nolt, J. (2015). Casualties as a moral measure of climate change. *Climatic Change,* *130*(3), 347–358. https://doi.org/10.1007/s10584-014-1131-2

Nucci, L., & Turiel, E. (2009). Capturing the complexity of moral development and education. *Mind, Brain, and Education, 3*(3), 151–159. https://doi.org/10.1111/j.1751-228X.2009.01065.x

Ojala, M. (2015a). Young people and global climate change: Emotions, coping, and engagement in everyday life. In N. Ansell, N. Klocker & T. Skelton (Hrsg.), *Geographies of global issues: Change and threat* (p. 1–19). Springer Singapore. https://doi.org/10.1007/978-981-4585-95-8_3-1

Ojala, M. (2015b). Hope in the face of climate change: Associations with environmental engagement and student perceptions of teachers' emotion communication style and future orientation. *The Journal of Environmental Education, 46*(3), 133–148. https://doi.org/10.1080/00958964.2015.1021662

Ojala, M. (2016). Preparing children for the emotional challenges of climate change: A review of the research. In K. Winograd (Ed.), *Education in times of environmental crises: Teaching children to be agents of change* (pp. 210–218). Routledge.

Ojala, M., & Anniko, M. (2020). Climate change as an existential challenge: Exploring how emerging adults cope with ambivalence about climate-friendly food choices. *Psyke & Logos, 41,* 17–33.

Onwezen, M. C., Antonides, G., & Bartels, J. (2013). The norm activation model: An exploration of the functions of anticipated pride and guilt in pro-environmental behaviour. *Journal of Economic Psychology, 39,* 141–153. https://doi.org/10.1016/j.joep.2013.07.005

Pauls, I. L., Shuman, E., Zomeren, M., Saguy, T., & Halperin, E. (2022). Does crossing a moral line justify collective means? Explaining how a perceived moral violation triggers normative and nonnormative forms of collective action. *European Journal of Social Psychology, 52*(1), 105–123. https://doi.org/10.1002/ejsp.2818

Pearson, A. R., Tsai, C. G., & Clayton, S. (2021). Ethics, morality, and the psychology of climate justice. *Current Opinion in Psychology, 42,* 36–42. https://doi.org/10.1016/j.copsyc.2021.03.001

Pihkala, P. (2020). Anxiety and the ecological crisis: An analysis of eco-anxiety and climate anxiety. *Sustainability, 12*(19), 7836. https://doi.org/10.3390/su12197836

Rhee, J. J., Schein, C., & Bastian, B. (2019). The what, how, and why of moralization: A review of current definitions, methods, and evidence in moralization research. *Social and Personality Psychology Compass, 13*(12). https://doi.org/10.1111/spc3.12511

Rodney, P. A. (2017). What we know about moral distress. *The American Journal of Nursing, 117*(2), S7–S10.

Rozin, P. (1999). The process of moralization. *Psychological Science, 10*(3), 218–221. https://doi.org/10.1111/1467-9280.00139

Ryan, T. J. (2019). Actions versus consequences in political arguments: Insights from moral psychology. *The Journal of Politics, 81*(2), 426–440. https://doi.org/10.1086/701494

Sanson, A. V., Burke, S. E. L., & Van Hoorn, J. (2018). Climate change: Implications for parents and parenting. *Parenting, 18*(3), 200–217. https://doi.org/10.1080/15295192.2018.1465307

Schneider, C. R., Zaval, L., Weber, E. U., & Markowitz, E. M. (2017). The influence of anticipated pride and guilt on pro-environmental decision making. *PLoS ONE, 12*(11), e0188781. https://doi.org/10.1371/journal.pone.0188781

Serfioti, D., Murphy, D., Greenberg, N., & Williamson, V. (2023). Professionals' perspectives on relevant approaches to psychological care in moral injury: A qualitative study. *Journal of Clinical Psychology, 79*(10), 2404–2421. https://doi.org/10.1002/jclp.23556

Shay, J. (2014). Moral injury. *Psychoanalytic Psychology, 31*(2), 182–191. https://doi.org/10.1037/a0036090

Singer, P. (2006). Ethics and climate change: A commentary on MacCracken, Toman and Gardiner. *Environmental Values, 15*(3), 415–422.

Skitka, L. J. (2010). The psychology of moral conviction. *Social and Personality Psychology Compass, 4*(4), 267–281. https://doi.org/10.1111/j.1751-9004.2010.00254.x

Skitka, L. J., Hanson, B. E., Morgan, G. S., & Wisneski, D. C. (2021). The psychology of moral conviction. *Annual Review of Psychology, 72*(1), 347–366. https://doi.org/10.1146/annurev-psych-063020-030612

Solmi, M., Radua, J., Olivola, M., Croce, E., Soardo, L., Salazar De Pablo, G., Il Shin, J., Kirkbride, J. B., Jones, P., Kim, J. H., Kim, J. Y., Carvalho, A. F., Seeman, M. V., Correll, C. U., & Fusar-Poli, P. (2022). Age at onset of mental disorders worldwide: Large-scale meta-analysis of 192 epidemiological studies. *Molecular Psychiatry, 27*(1), 281–295. https://doi.org/10.1038/s41380-021-01161-7

Sternäng, L., & Lundholm, C. (2011). Climate change and morality: Students' perspectives on the individual and society. *International Journal of Science Education, 33*(8), 1131–1148. https://doi.org/10.1080/09500693.2010.503765

Ten Have, H., & Patrão Neves, M. D. C. (2021a). Moral distress. In *Dictionary of global bioethics* (p. 731). Springer International Publishing. https://doi.org/10.1007/978-3-030-54161-3

Ten Have, H., & Patrão Neves, M. D. C. (2021b). Moral residue. In *Dictionary of global bioethics* (p. 743). Springer International Publishing. https://doi.org/10.1007/978-3-030-54161-3

Ter Heide, F. J. J., & Olff, M. (2023). Widening the scope: Defining and treating moral injury in diverse populations. *European Journal of Psychotraumatology, 14*(2), 2196899. https://doi.org/10.1080/20008066.2023.2196899

Thunberg, G. (2019, September 23). *Greta Thunberg's speech at the U.N. climate action summit* [Speech transcript]. www.npr.org/2019/09/23/763452863/transcript-greta-thunbergs-speech-at-the-u-n-climate-action-summit

Trott, C. D. (2022). Climate change education for transformation: Exploring the affective and attitudinal dimensions of children's learning and action. *Environmental Education Research, 28*(7), 1023–1042. https://doi.org/10.1080/13504622.2021.2007223

Vamvalis, M. (2023). "We're fighting for our lives": Centering affective, collective and systemic approaches to climate justice education as a youth mental health imperative. *Research in Education, 117*(1), 88–112. https://doi.org/10.1177/00345237231160090

Vergunst, F., & Berry, H. L. (2022). Climate change and children's mental health: A developmental perspective. *Clinical Psychological Science, 10*(4), 767–785. https://doi.org/10.1177/21677026211040787

Weintrobe, S. (2020). Moral injury, the culture of uncare and the climate bubble. *Journal of Social Work Practice, 34*(4), 351–362. https://doi.org/10.1080/02650533.2020.1844167

5 Solutions for Climate-Conflicted Children of the Global North

Tim Kelly

Positionality Statement

I have written this chapter from the perspective of a high school teacher (Science, Agriculture, and Psychology) and a researcher in environmental and educational psychology. I acknowledge that I am Pākehā (white European) who has a strong appreciation for the Māori concept of *kaitiakitanga* (guardianship of the land). For the past 20 years, I have lived and worked in a small rural New Zealand farming community, which typically votes against climate change policies. I, however, am a firm believer that the importance of climate action must be ingrained in children before their personal values become too fixed, as it is the voters of the future who will determine whether or not we can reduce the deleterious impact of human actions on the climate.

Introduction

Why are children anxious about climate change? To answer this, let us consider the lives of three children: Aanya, Sam, and Sasha. These children, while fictional, are composites of children I have interviewed or whose interviews I have read.

For some children, especially those in the Global South, climate change is a significant threat to their survival, and an inescapable source of anxiety:

> Sometimes, when we can afford it, I go to school. I have not been this year. Our village was flooded and we lost our crops. The people in our village managed to save most of the seed we had stored, and we will plant again this season. But we have nothing to sell now, and I worry that the floods will come again.
>
> (Aanya, Bangladesh)

While underprivileged children like Aanya justifiably fear climate change, well-off children may also fear that their survival is threatened by climate change, but irrationally so. These children tend to fear an ambiguous global threat that could destroy all life on Earth (e.g. "humanity is doomed" (Hickman et al., 2021)),

DOI: 10.4324/9781003494416-6

while the reality is that the wealthy have the ways and means to avoid the worst effects of climate change (Sanson & Bellemo, 2021). This unfounded anxiety related to personal security can be resolved through educating children about the realistic outcomes of climate change in the Global North, while taking care not to skirt around the fact that climate change will have devastating effects for other species and other people, particularly the poor of the Global South.

When children in the Global North become aware that the negative outcomes of climate change are in part caused by their own personal behaviours but are unlikely to significantly threaten their own security, they may experience more subtle types of anxiety, which are the focus of this chapter:

At school we learn about climate change and how our behaviours affect other people and other species. It makes me think about the impact of my lifestyle on the planet, but everyone I know has a similar lifestyle to me. We all use a lot of electricity, we drive everywhere we need to go, we waste food and other resources. I feel bad about it, but at the same time I feel that the way I am is normal. I think it would be less stressful to just go with the flow and stop worrying about my impact on the climate.

(Sam, New Zealand)

I'd like to buy a car when I turn seventeen next year, but part of me feels like I should stick to biking or taking the bus, to reduce my impact on the climate.

(Sasha, Ireland)

Sam feels conflicted when his behaviour is harmful to the climate; he feels that it is morally wrong, but his behaviour reflects what is normal in society. Sam's low-level anxiety is highly important because of how he chooses to manage it: if he chooses to reduce his anxiety by ignoring his conscience and continuing to behave as others do, this, I am sure you will be aware, is the recipe for Aanya's own anxiety.

Sasha also feels conflicted but for different reasons. Sasha wants to buy a car, presumably for the convenience or pleasure it would give her, but her conscience tells her that it is morally wrong to consume fossil fuels. Sasha is conflicted between acting for her own benefit (self-enhancement) or acting in a manner that does not impact negatively on the lives of others (self-transcendence). As with Sam, the way in which Sasha chooses to resolve this conflict can have negative repercussions for vulnerable people and species.

Considering these examples, in this chapter we will explore the anxiety experienced by children who, on the one hand, are told by teachers and the media that they are responsible for the negative outcomes of climate change, while on the other hand, they are trying to fit into a society in which climate-harmful behaviour is normalized. In addition, we will investigate the anxiety that children can feel when their differing personal values conflict, making behaviour choices worrisome. Although challenging, solutions to both these sources of anxiety are possible, and we will delve into some individual- and group-level resolutions to the described conflicts.

Cognitive Dissonance

This chapter is not concerned with anxiety that results from a fear for one's own personal security, or the security of one's significant others. Instead, this chapter deals with the mental conflict that arises for children in the Global North when (a) there is a conflict between what they believe is right and what is clearly normal in society, and/or (b) they experience a conflict between their own differing values. This concept of mental conflict, and the anxiety that arises from it, needs to be explored before discussing society's norms or individuals' personal values, and is best explained by Festinger's (1957) Theory of Cognitive Dissonance. Cognitive dissonance refers to the discomfort or tension that arises when an individual's behaviour conflicts with their beliefs, or when they hold two or more conflicting beliefs, attitudes, or values. The term "cognitive dissonance" implies a state of mental discomfort or inconsistency, and the theory posits that individuals are motivated to resolve this inconsistency and restore a sense of coherence. The conflict can be resolved if the person changes their behaviour, values, beliefs, or attitudes to make them more congruent with each other. However, if the dissonance remains unresolved, it can manifest in an underlying anxiety. A person's behaviour, values, beliefs, and attitudes form their identity, and changing these can be both challenging and undesirable, to the degree that low-level anxiety is preferable to change. In addition, cognitive dissonance is often related to behaviours that we feel obliged to undertake. Employment is a common example: if a person feels that their job is a drudge or is stressful, it will cause cognitive dissonance until they can either find a way to get satisfaction from the job or change jobs. Given that society places many demands on us, and that we must regularly make conflicting choices, it is common for people to experience cognitive dissonance, often chronically.

As you read through this chapter, consider how climate change might cause cognitive dissonance for the children you know. Are any of these children anxious about the impact of their own behaviour, or only the behaviour of others? Why are some of the children you know not concerned about the impact of climate change? To help clarify our thoughts about these children, let us first explore what is considered normal behaviour in the Global North.

What Is Normal?

What is normal in the societies of the wealthy countries we call the "Global North"? In the Global North, consumerism is a prevalent norm. The culture of the Global North encourages the constant pursuit of new products and the disposal of older ones, contributing to resource depletion and waste generation. High levels of resource consumption, such as energy, water, and materials, are often considered normal, especially in more affluent communities. Reliance on private vehicles is widespread in the Global North, contributing significantly to carbon emissions. Car ownership and the preference for personal transportation over public transit are often deeply ingrained behaviours. High energy consumption, often powered by the use of fossil fuels, is a common feature in many households and industries. Diets in the Global North tend to include a high proportion of resource-intensive and environmentally damaging foods, such as meat and dairy. The demand for these products contributes to deforestation, methane emissions, and other environmental issues. Disposable and single-use products are widely accepted, contributing to significant amounts of waste. Recycling and waste reduction efforts may vary, but in most cases, there is room for improvement in minimizing the environmental impact of consumption. While there is a growing awareness of environmental issues, the widespread adoption of eco-friendly behaviours is inhibited by various factors.

Norms

For a child growing up in the Global North, the practices earlier are evident in their day-to-day lives and are considered normal. These are the behaviours of friends, of family members, of teachers, of media and sports personalities, of politicians, and of most other people in society. A small minority of people have rejected these behaviours, but the *descriptive norm* is one of over-consumption and waste. Descriptive norms indicate what is normal in society and are important determinants of behaviour (Cialdini et al., 1990; Schultz et al., 2007; Smith et al., 2012; White et al., 2009). Descriptive norms also influence what we consider to be important, and if an individual perceives that a behaviour is not widely adopted, it can decrease the likelihood that the individual will adopt that behaviour (Smith et al., 2012). Indeed, descriptive norms can play a significant role in the development or reinforcement of *personal norms* (an individual's internalized standards for behaviour). If an individual observes certain behaviours as common or socially accepted (descriptive norm), it may influence their internalization of those behaviours as personally acceptable or unacceptable (personal norm).

In the case of Sam and Sasha, two of the children introduced at the start of this chapter, the descriptive norm in their societies (New Zealand and Ireland) is very much climate-harmful. However, Sam and Sasha's teachers and other influences have caused them to develop a climate-conscious personal norm, such that they care about their impact on the climate. Sasha and Sam's climate-conscious personal norms cause them anxiety, whereas their peers who do not care about the effects of climate change on other people and species do not experience that anxiety. Why did Sam and Sasha develop this seemingly adverse (for them) personal norm, whereas many of their classmates did not?

What Causes a Climate-Conscious Personal Norm?

Personal norms guide an individual's choices and actions, and they do not always align with the descriptive norms observed in that society. Personal norms are influenced by a combination of factors, including personal values, personal experiences, moral upbringing, and internal moral reasoning. Even when the rest of society seems to believe in or practice something different, a person can still develop an independent personal norm. Going forward, we will focus specifically on the influence of children's personal values on the formation of climate-friendly personal norms.

Values

If personal norms are an individual's internalized standards for behaviour, then what are personal values? Personal values, also known as *basic human values*, serve as stable, broad, guiding principles that shape individuals' attitudes, beliefs, and behaviours (Schwartz, 1994, 2017). Unlike personal norms, they do not relate to specific behaviours, but rather to a collective of behaviours and priorities. Indeed, the purpose of values is to reduce the cognitive load we experience when making behaviour decisions, by defaulting to whatever behaviour fits best with our values. For example, if a person has strong ecological values, they are more likely to bin a piece of litter rather than drop it on the ground. A person who strongly values benevolence is likely to be loyal to their friends, even if it causes hardship for themselves. A person who strongly values power may prioritize becoming wealthy over aspects of life that other people might consider important.

It is important to note that our values are not always reflected in our behaviours, as sometimes other factors are more pressing, especially if the value is not strongly held. Values Beliefs Norms (VBN) theory (Stern, 2000) indicates that values are the foundation for our pro-environmental behaviours, and that our behaviour intentions are further refined by our specific beliefs and personal norms. As such, in the case of littering, a person who does not value the natural environment strongly is more likely to litter than a person who does. However, valuing the natural environment does not guarantee that the person will not litter. In addition to valuing the natural environment, the person must believe that littering will have an adverse effect on the environment (beliefs) and must also believe that littering is wrong (personal norm). Research also indicates that descriptive norms have an influence on these behaviour decisions (Collado et al., 2019; Smith et al., 2012). For example, if I have strong ecological values and I believe that littering is harmful and wrong, will I litter if there is already litter all over the ground and most other people seem to be littering? It is not uncommon, after all, for people to compromise their own values, when their values do not appear to be the norm. When we consider the relationship between a person's values and their climate-impacting behaviours, we need to consider values other than just their ecological values. Empirical research by Shalom Schwartz and others (e.g. Bilsky et al., 2011; Cieciuch & Schwartz, 2012; Schwartz, 1992, 1994, 2017; Schwartz et al., 2001, 2017) has found that basic human values can be placed on a continuum, with the result that some groups of values oppose other groups of values. For example, values that reflect openness to change are not congruent with values that reflect conservatism. In this example, a person who values conservatism more than openness to change is more likely to behave in a manner that is conservative. In addition to openness to change values and conservatism values, the two other primary groupings of values on Schwartz's basic human values continuum are self-transcendence values and self-enhancement values. Self-transcendence values and self-enhancement values are particularly relevant to the discussion of climate anxiety in the Global North.

Schwartz and colleagues have found that people who have strong ecological values tend to also have strong altruism values, and these values are together grouped as *self-transcendence values*. Self-transcendence values emphasize the well-being of others, nature, and the collective good. These values encompass empathy, social justice, environmental preservation, and the pursuit of a sustainable future. Ecological values cause our concern for the continued survival of other species in the face of climate change, and altruistic values cause our concern for the welfare of other people, in this case those people most vulnerable to the impacts of climate change. When we teach children about the importance of reducing our impact on the climate, we should be trying to tap into their self-transcendence values.

On Schwartz's continuum of basic human values, *self-enhancement values* oppose self-transcendence values. Self-enhancement values prioritize power, wealth, and achievement. They emphasize personal success, status, and self-interest. Modern environmental research also categorizes hedonic values, which prioritize convenience and pleasure, as self-enhancing values (Steg et al., 2014). Self-enhancement values are highly correlated with the concept of materialistic values, which forms part of other theories of environmental behaviour (Hurst et al., 2013). Materialistic values, as the name suggests, concern the prioritization of goods and wealth.

It is possible for a person to value both self-transcendence and self-enhancement, and most people do. However, people ultimately prioritize one over the other, and this affects their decision making. In the context of climate change, self-enhancement values can cause individuals to place their own immediate desires above the long-term well-being of other people and species (De Groot & Steg, 2010). The relative strength of the opposing values is key. For example, a person with strong self-transcendence values not only may believe that it is important to reduce greenhouse gas emissions (self-transcendence) but also want to travel overseas on holiday (self-enhancement). Taking an overseas holiday will result in substantial greenhouse emissions, so how does the person react? If her self-transcendence values are genuinely stronger than her self-enhancement values, we would expect her to not travel by airplane on holiday. So why is a discussion about basic human values important in a book about climate change anxiety? If we go back to Sasha, the Irish girl introduced at the start of the chapter, you may remember that she was conflicted between buying a car or not. Coming of age to be able to buy a car is likely to be a significant milestone in Sasha's life, but she feels that it is wrong to contribute more greenhouse gases to the atmosphere through driving. This conflict causes cognitive dissonance for Sasha and low-level anxiety. Whether Sasha decides to buy a car or not will depend in a large part on the strength of her self-enhancement values when compared to her self-transcendence values. If Sasha does buy a car, she will try to rationalize this action to reduce her cognitive dissonance. If Sasha does not buy the car, she is likely to feel like she is missing out, which itself is a form of cognitive dissonance. How does Sasha reduce this secondary cognitive dissonance that can result from doing what is morally right? How do people rationalize the inconvenience and loss of materialistic self-enhancement that is part and parcel of many environmentally friendly behaviours?

Reducing Cognitive Dissonance and Associated Anxiety

So far we have explored two potential causes of cognitive dissonance: Sam from New Zealand experiences cognitive dissonance because even though he feels like he should act in a climate-friendly way, doing so does not seem to be normal in his society. Sasha from Ireland experiences cognitive dissonance because her self-enhancement values are conflicting with her self-transcendence values; she would like to buy a car, but doing so would increase greenhouse gas emissions. These dilemmas are important for educators to understand because we cannot realistically expect all children to persist with optional behaviours that cause them cognitive dissonance. The easiest way for children to reduce their cognitive dissonance in this context is to reduce the importance they ascribe to the impacts of climate change and not bother with pro-environmental behaviours, but this is obviously the worst-case outcome for environmental educators and the environment. In that case, how do we help children overcome the cognitive dissonance that accompanies acting on their self-transcendence values?

In the societies of the Global North, possessing and acting on strong self-transcendence values is less common than we would like to believe. Being environmentally friendly is generally considered to be a pro-social characteristic and so people like to think of themselves as pro-environmental and undertake small gestures which enable them to feel that they are doing their part for the environment, such as reusing shopping bags or recycling. The reality is that acting in a climate-friendly manner often requires us to compromise our self-enhancement values, such that we forego activities that are pleasurable, such as eating red meat, or undertaking activities in a way that is less convenient, such as taking public transport instead of the car. Having said that, small gestures can raise awareness of an issue in society, and maybe this is enough to build momentum to influence the most important pro-environmental behaviour of all: voting for a pro-environmental government. Although children cannot vote, the values they develop in childhood will carry through to adulthood and influence their voting choices.

If we think back to the described norms of the Global North listed earlier in this chapter, it is clear that the listed behaviours are more congruent with self-enhancement than self-transcendence values, with respect to their impact on climate change. If society seems to predominantly act in a way that is climate-harmful, what is the risk in encouraging children to adopt climate-friendly behaviours that seemingly go against the grain?

One of the most important *basic human needs* that people have, especially children, is the need to feel *relatedness* to other people (Ryan & Deci, 2000). When children voluntarily change their behaviour to better reflect their values and attitudes, this behaviour shift can be problematic if the behaviour does not align with the descriptive norm in society. For children, it can be very important to fit in with their peers, and so a child may be reluctant to act in a way that is significantly different from their peers (Harter et al., 1996). How can we help these children cross this barrier and persist with behaviour that is environmentally positive, but that potentially differentiates them from their peers, family members, and the other people who are significant in their lives?

Individual-Level Solutions

There are two primary solutions to the conflict a child experiences when he wishes to go against the grain: the first solution is to not go against the grain, but this is obviously not desirable from the environmental educator's perspective. The second solution is to change the "grain," that is, for the significant others in the child's life to have similar values and behaviours to the child. This can occur by selection, in the case of friends, whereby the child associates with peers that he knows share similar values to himself (Brechwald & Prinstein, 2011). This can be facilitated in schools through creating environment clubs or similar, where like-minded students can interact with each other. These groups tend to focus on direct action, such as planting trees, or indirect action, such as influencing other people and policy makers. Being part of a group helps reduce the child's feeling that their behaviour is not normal. An alternative solution is for the people in the child's life to make concessions to adapt their own behaviour to better align with the child's behaviour. For example, if a child in a family feels conflicted about eating red meat because of its impact on the climate, the family can support the child by reducing how much red meat they eat, and verbalizing that they are doing so because they share similar concerns to the child. In the classroom, the teacher can create student-centred systems that emphasize the reduction of waste, such as giving students control of managing the air conditioning and heating. It is important to put time into making these processes the descriptive norm and to teach students about the importance of the action. When choosing proponents of a student-led environmental initiative, teachers should ensure that an influential member from each social group within the class is given a role of responsibility; if these students become enthusiastic about the initiative, they are more likely than randomly selected students to have an influence on their peers. In such ways, it is possible to make the classroom descriptive norm more climate-friendly, which in turn helps students feel more positive about their pro-environmental behaviours.

There is another approach to changing the descriptive norm, which is the basis of recent research: that is to change the child's perception of what the people in his life think is important. Children tend to underestimate how much their peers value the natural environment (Kelly et al., 2023; Long et al., 2014), and thus if children come to understand what is actually important to their peers, they may change their opinion of what is normal in society and become more willing to persist with climate-friendly behaviour. In addition to providing a supportive environment, enabling group activities in pro-environmental contexts is a useful way to help children become more aware of their peers' values. When children work with other children on environmental projects or undertake sustained environmental education activities that require them to share their opinions with their peers, they can more easily see how important environmentalism is to their peers. These group activities are likely to be particularly useful if they involve environmental dialogue between the participants and lead to the participants expressing their own values, even indirectly. Research indicates that when children engage in environmental dialogue with each other, this leads them to become more likely to engage in further environmental dialogue in the longer term (Kelly et al., 2023). We assume that when children feel comfortable speaking to each other about environmental matters, this allows them to feel that their own values and attitudes are normal, which in turn reduces the cognitive dissonance they may experience when they feel like their self-transcendence values and behaviours are not the norm.

Let us return to our New Zealand case study, Sam, who wanted to act in a climate-friendly way but felt that the effort he put into such behaviours may not be worth it, as it appeared to him that it was not normal to behave in such a way. Sam was reluctant to talk about these feelings with his significant others, but his teacher encouraged him to present his feelings in a speech competition at school. Sam's parents were there when he gave his speech. Sam's parents shared many of Sam's concerns about climate change, and Sam's speech was the impetus they needed to make some climate-friendly changes in their lives. Sam's family has since replaced their SUV with an electric car. They now prepare family meals containing red meat no more than once a week and are supportive of Sam foregoing red meat altogether. They have decided to spend their holidays locally this year instead of going overseas – after all, who wants to go overseas when you live in New Zealand! Since making his speech, Sam has noticed that people are more likely to talk with him about environmental matters and about the things they do to help the environment. Although Sam still feels like his climate-friendly behaviours are uncommon in society, he does feel more relatedness in this context to the people who are closest to him. Sam says that it is true to say that he now feels less conflict between his choice to act in a climate-friendly way and his desire to fit in with society.

Group-Level Solutions

Although individual-level solutions can be suggested, are there group-level solutions to the cognitive dissonance that children may feel when their desire to be climate-friendly conflicts with what they encounter in society? It is important to acknowledge that societal-level changes are slow unless society is in clear and present danger. Many people believe that societies in the Global North will not take significant action to curb global warming until the consequences of climate change result in loss for more people and have an economic impact on industry and governments (Frantz & Mayer, 2009), and for that reason societal-level changes to counter global warming are likely to be slow. Nonetheless, there are some societal-level changes that are we beginning to see that have the potential to change the descriptive norm from climate-harmful to climate-friendly.

Convenience and cost are important to people in the Global North. If a behaviour can be adapted to become more climate-friendly, without becoming significantly more inconvenient or expensive, then people are likely to adopt that behaviour into society, because it is considered pro-social. Recycling is a case in point. In most Global North societies, recycling is now the descriptive norm for waste management. Children who recycle their waste have no reason to feel that they are behaving abnormally, and the behaviour is unlikely to cause cognitive dissonance as a result. While the adoption of recycling is a positive step that shows societal-level change can occur, it is a relatively minor behaviour in the scheme of things. More significant measures require the majority of society to support their implementation through government policy. When a society votes for a government that has a pro-environment agenda, significant changes can occur. An example in Sam's home country, New Zealand, is the tax rebate that was introduced for electric vehicles by a pro-environmental government in 2021, and then removed by the subsequent less pro-environmental government in 2024. In Sasha's home country, Ireland, new houses must be built with a high degree of energy efficiency, which usually includes the installation of solar panels. Both of these visible society-wide measures help frame the descriptive norm for children like Sam and Sasha.

New technology is often seen as the primary way to shift societies in the Global North from climate-harmful to climate-friendly behaviours (Barrett, 2009). The aim of this technology is often to modify behaviours to make them less climate-harmful while retaining their convenience and self-enhancement aspects. As such, technological improvements do not require people to change their values and may be a more realistic way to reduce our impact on the climate (Manfredo et al., 2017). As George Bush Sr famously informed the public at the 1992 Earth Summit, "The American way of life is not up for negotiations. Period" (Deen, 2012).

The uptake of technology tends to be driven not just by wants and needs but also by expense. Electric vehicles and solar panels are cases in point: as these technologies have become more affordable than their climate-harmful predecessors, their uptake has been relatively swift. Affordable climate-friendly technology is likely to be one of the most significant factors in changing the descriptive norm. When children encounter increasingly more electric vehicles on the street, and see solar panels on buildings wherever they go, this sends a message that society is putting its words into action and that climate-friendly behaviour is not unusual. These progressive changes to the descriptive norm help reduce the cognitive dissonance that children like Sam experience when they feel like their own efforts are not mirrored in society.

For children like Sam and Sasha, the media can also play a large part in framing the descriptive norm. The media they are exposed to can reinforce either a negative or positive descriptive norm for environmental behaviours. Sam listens to rap music and finds that the lyrics usually promote self-enhancement, especially the prioritization of individual wealth and power. Sasha reads the *Guardian* newspaper, a left-wing publication that has a daily focus on climate change issues. Both Sam and Sasha read climate change content on social media and find that such content is "pushed" in their media channels as a result (see Swart (2021) for a review of young people's interactions with algorithmic news selection). Although social media algorithms tend to get a bad rap (De Vito et al., 2017; Etter & Albu, 2021), their usefulness for promoting pro-environmental content to those who require affirmation of their climate-friendly values and behaviours is substantial.

Despite the advances noted earlier, if the descriptive norm in the Global North continues to be primarily climate-harmful, how do children reconcile this with their own values and behaviours? The answer perhaps lies in identity psychology. People who have strong self-transcendence values can develop a strong *climate-friendly identity*, such that being climate-friendly is an important part of who they are (Gatersleben et al., 2014; van der Werff et al., 2013). Having a strong identity helps build resilience in the face of conflicting motivations. For example, when a professional football team is doing badly, which fans come back week after week to support them? Those fans whose identity is built around the football club. Similarly, building a strong climate-friendly identity can help children become less concerned with what everyone else is doing and to focus on doing what they themselves believe is important. Building a strong identity is helped by being part of a group of similar people (Terry et al., 1999), and so enabling children to work together on pro-environmental projects can help grow their climate-friendly identity.

Sasha's conflict is related to the idea that sacrifices need to be made to aspects of our lifestyle for the sake of a stable climate. Sasha wants to buy a car but worries about her potential impact on global warming. One approach to lessening her concern is to suggest that she buy a car that has low environmental impact, in which case her self-enhancement values are unlikely to be compromised, provided the car is affordable. The other approach is for Sasha to forego her self-enhancement and not buy the car. To do so will require strong self-transcendence values and a strong climate-friendly identity. The stronger Sasha's identity in this regard, the less likely she is to experience cognitive dissonance related to her decision.

The ultimate solution to the cognitive dissonance we have explored in this chapter is for society as a whole to become more climate-friendly. This may occur through advances in technology and decreases in the cost of climate-friendly alternatives. Can we hope to change society's values? Some researchers argue that it is naive to hope for societies in the Global North to change their values for the sake of the natural environment (Manfredo et al., 2017) and instead believe that changes in technology and expense are the most viable solutions. Nonetheless, is it possible to change the values within a smaller section of a society, such as a school? School is where children spend a large part of their lives, and the culture in the school can influence their cognitive and moral development. While it seems plausible that the whole-school environment could influence children's ecological values, research suggests that this may not be the case. Pauw and Van Petegem (2013) found no difference between the pro-environmental behaviours and ecological (preservation) values of children who attended either eco-schools or normal schools in Belgium. This raises the issue of how difficult it is to influence a person's values, especially when we consider that personal values are strongly held beliefs.

Can Values Be Changed?

Our personal values are very stable (Vecchione et al., 2016), and by adulthood they become very difficult to change. Changing one's values means that one must also change a whole suite of behaviours and attitudes in response, which is something that would require great motivation and determination. Indeed, a change in one's values usually means a change in one's identity. However, it is during childhood that a person's values are most malleable (Vecchione et al., 2020), and so childhood is the critical time for environmental educators to engender strong self-transcendence values in people. In early adolescence in particular, children begin to seek to create their own personal identity, and begin to look for other sources of identity, such as peers, media, etc. (Harter et al., 1996). If we hope to create societal-level changes in people's concern regarding climate change, we must begin by influencing our youth to embrace self-transcendence, even though our own generation is a poor role model for such values. Teachers are some of the most significant people in a child's life, and as such they have a great responsibility to not just educate students about the importance of self-transcendence but also role model self-transcendence values and create opportunities for students to explore and express their own values with their peers. If children can see that behaving in a climate-friendly way is normal, then the cognitive dissonance and associated anxiety they can experience when being asked to act in such a way can be minimized.

Conclusion

Climate change threatens the survival of many people and species. Although children in the Global North are less susceptible to the threats of climate change, it is morally imperative that they should develop concern for the welfare of the vulnerable people and species who will be affected by the climate changing behaviour of societies in the Global North. However, with concern and the awareness of responsibility comes the likelihood of cognitive dissonance and the low-level anxiety that accompanies it. Educators need to be aware that it is easiest for children to reduce this cognitive dissonance by downplaying the seriousness of climate change and their role in creating such change. As such, teachers and other people of influence need to create an environment wherein acting in a climate-friendly manner is considered the norm, because when children feel normal, the likelihood that they will experience cognitive dissonance is reduced. Similarly, if we can engineer ways for students to build a climate-friendly identity, sacrificing some self-enhancement for the sake of others will feel less conflicting. Whereas many teachers have embraced teaching children about their climate change responsibilities, their critical role is helping children adapt to the weight of this responsibility.

References

Barrett, S. (2009). The coming global climate-technology revolution. *Source: The Journal of Economic Perspectives, 23*(2).

Bilsky, W., Janik, M., & Schwartz, S. H. (2011). The structural organization of human values-evidence from three rounds of the European social survey (ess). *Journal of Cross-Cultural Psychology, 42*(5), 759–776. https://doi.org/10.1177/0022022110362757

Brechwald, W. A., & Prinstein, M. J. (2011). Beyond homophily: A decade of advances in understanding peer influence processes. *Journal of Research on Adolescence, 21*(1), 166–179. https://doi.org/10.1111/j.1532-7795.2010.00721.x

Cialdini, R. B., Reno, R. R., & Kallgren, C. A. (1990). A focus theory of normative conduct: Recycling the concept of norms to reduce littering in public places. *Journal of Personality and Social Psychology, 58*(6), 1015–1026. https://doi.org/10.1037/0022-3514.58.6.1015

Cieciuch, J., & Schwartz, S. H. (2012). The number of distinct basic values and their structure assessed by PVQ-40. *Journal of Personality Assessment, 94*(3), 321–328. https://doi.org/10.1080/00223891.2012.655817

Collado, S., Staats, H., & Sancho, P. (2019). Normative influences on adolescents' self-reported pro-environmental behaviors: The role of parents and friends. *Environment and Behavior, 51*(3), 288–314. https://doi.org/10.1177/0013916517744591

Deen, T. (2012). *U.S. lifestyle is not up for negotiation.* Inter Press Service News Agency. www.ipsnews.net/2012/05/us-lifestyle-is-not-up-for-negotiation/

De Groot, J. I. M., & Steg, L. (2010). Relationships between value orientations, self-determined motivational types and pro-environmental behavioural intentions. *Journal of Environmental Psychology, 30*(4), 368–378. https://doi.org/10.1016/j.jenvp.2010.04.002

De Vito, M. A., Gergle, D., & Birnholtz, J. (2017). "Algorithms ruin everything": #RIPTwitter, folk theories, and resistance to algorithmic change in social media. *Conference on human factors in computing systems – proceedings, 2017-May* (pp. 3163–3174). https://doi.org/10.1145/3025453.3025659

Etter, M., & Albu, O. B. (2021). Activists in the dark: Social media algorithms and collective action in two social movement organizations. *Organization, 28*(1), 68–91. https://doi.org/10.1177/1350508420961532

Festinger, L. (1957). A theory of cognitive dissonance. In *A theory of cognitive dissonance.* Stanford University Press.

Frantz, C. M., & Mayer, F. S. (2009). The emergency of climate change: Why are we failing to take action? *Analyses of Social Issues and Public Policy, 9*(1), 205–222. https://doi.org/10.1111/j.1530-2415.2009.01180.x

Gatersleben, B., Murtagh, N., & Abrahamse, W. (2014). Values, identity and pro-environmental behaviour. *Contemporary Social Science, 9*(4), 374–392. https://doi.org/10.1080/21582041.2012.682086

Harter, S., Stocker, C., & Robinson, N. S. (1996). The perceived directionality of the link between approval and self-worth: The liabilities of a looking glass self-orientation among young adolescents. *Journal of Research on Adolescence, 6*(3), 285–308.

Hickman, C., Marks, E., Pihkala, P., Clayton, S., Lewandowski, R. E., Mayall, E. E., Wray, B., Mellor, C., & van Susteren, L. (2021). Climate anxiety in children and young people and their beliefs about government responses to climate change: A global survey. *The Lancet Planetary Health, 5*(12), e863–e873. https://doi.org/10.1016/S2542-5196(21)00278-3

Hurst, M., Dittmar, H., Bond, R., & Kasser, T. (2013). The relationship between materialistic values and environmental attitudes and behaviors: A meta-analysis. *Journal of Environmental Psychology, 36*, 257–269. https://doi.org/10.1016/j.jenvp.2013.09.003

Kelly, T., Bouman, T., Kemp, S., Wijngaarden, F., & Grace, R. C. (2023). Exploration of children's value patterns in relation to environmental education programmes. *Frontiers in Psychology, 14*. https://doi.org/10.3389/fpsyg.2023.1264487

Long, J., Harré, N., & Atkinson, Q. D. (2014). Understanding change in recycling and littering behavior across a school social network. *American Journal of Community Psychology, 53*(3–4), 462–474. https://doi.org/10.1007/s10464-013-9613-3

Manfredo, M. J., Bruskotter, J. T., Teel, T. L., Fulton, D., Schwartz, S. H., Arlinghaus, R., Oishi, S., Uskul, A. K., Redford, K., Kitayama, S., & Sullivan, L. (2017). Why social values cannot be changed for the sake of conservation. *Conservation Biology, 31*(4), 772–780. https://doi.org/10.1111/cobi.12855

Pauw, J. B. De, & Van Petegem, P. (2013). The effect of eco-schools on childrens environmental values and behaviour. *Journal of Biological Education, 47*(2), 96–103. https://doi.org/10.1080/00219266.2013.764342

Reser, J. P., Bradley, G. L., Glendon, A. I., Ellul, M. C., & Callaghan, R. (2012). *Public risk perceptions, understandings and responses to climate change and natural disasters in Australia, 2010 and 2011* (p. 246). National Climate Change Adaptation Research Facility. www.unisdr.org/preventionweb/files/30470_finalreportreserpublicriskperceptio.pdf

Ryan, R. M., & Deci, E. L. (2000). Intrinsic and extrinsic motivations: Classic definitions and new directions. *Contemporary Educational Psychology, 25*(1), 54–67. https://doi.org/10.1006/ceps.1999.1020

Sanson, A., & Bellemo, M. (2021). Children and youth in the climate crisis. *BJPsych Bulletin, 45*(4), 205–209. https://doi.org/10.1192/bjb.2021.16

Schultz, P. W., Nolan, J. M., Cialdini, R. B., Goldstein, N. J., & Griskevicius, V. (2007). The constructive, destructive, and reconstructive power of social norms. *Psychological Science, 18*(5), 429–434. https://doi.org/10.1111/j.1467-9280.2007.01917.x

Schwartz, S. H. (1992). Universals in the content and structure of values. *Advances in Experimental Social Psychology, 25*(C), 1–65. https://doi.org/10.1016/S0065-2601(08)60281-6

Schwartz, S. H. (1994). Are there universal aspects in the structure and contents of human values? *Journal of Social Issues.* http://doi.org/10.1111/j.1540-4560.1994.tb01196.x/abstract

Schwartz, S. H. (2017). The refined theory of basic values. In *Values and behavior* (pp. 51–72). Springer.

Schwartz, S. H., Cieciuch, J., Vecchione, M., Torres, C., Dirilen-gumus, O., & Butenko, T. (2017). Value tradeoffs propel and inhibit behavior: Validating the 19 refined values in four countries. *European Journal of Social Psychology, 47*, 241–258.

Schwartz, S. H., Melech, G., Lehmann, A., Burgess, S., Harris, M., & Owens, V. (2001). Extending the cross-cultural validity of the theory of basic human values with a different method of measurement. *Journal of Cross-Cultural Psychology, 32*(5), 519–542. https://doi.org/10.1177/0022022101032005001

Skeirytė, A., Krikštolaitis, R., & Liobikienė, G. (2022). The differences of climate change perception, responsibility and climate-friendly behavior among generations and the main determinants of youth's climate-friendly actions in the EU. *Journal of Environmental Management, 323*, 116277. https://doi.org/10.1016/j.jenvman.2022.116277

Smith, J. R., Louis, W. R., Terry, D. J., Greenaway, K. H., Clarke, M. R., & Cheng, X. (2012). Congruent or conflicted? The impact of injunctive and descriptive norms on environmental intentions. *Journal of Environmental Psychology, 32*(4), 353–361. https://doi.org/10.1016/j.jenvp.2012.06.001

Steg, L., Perlaviciute, G., van der Werff, E., & Lurvink, J. (2014). The significance of hedonic values for environmentally relevant attitudes, preferences, and actions. *Environment and Behavior, 46*(2), 163–192. https://doi.org/10.1177/0013916512454730

Stern, P. C. (2000). Toward a coherent theory of environmentally significant behavior. *Journal of Social Issues, 56*(3), 407–424. http://web.stanford.edu/~kcarmel/CC_BehavChange_Course/readings/Stern_metareview_2000.pdf

Swart, J. (2021). Experiencing algorithms: How young people understand, feel about, and engage with algorithmic news selection on social media. *Social Media and Society, 7*(2). https://doi.org/10.1177/20563051211008828

Terry, D. J., Hogg, M. A., & White, K. M. (1999). The theory of planned behaviour: Self-identity, social identity and group norms. *British Journal of Social Psychology, 38*, 225–244.

van der Werff, E., Steg, L., & Keizer, K. (2013). The value of environmental self-identity: The relationship between biospheric values, environmental self-identity and environmental preferences, intentions and behaviour. *Journal of Environmental Psychology*, *34*, 55–63. https://doi.org/10.1016/j.jenvp.2012.12.006

Vecchione, M., Schwartz, S. H., Alessandri, G., Döring, A. K., Castellani, V., & Caprara, M. G. (2016). Stability and change of basic personal values in early adulthood: An 8-year longitudinal study. *Journal of Research in Personality*, *63*, 111–122. https://doi.org/10.1016/J.JRP.2016.06.002

Vecchione, M., Schwartz, S. H., Davidov, E., Cieciuch, J., Alessandri, G., & Marsicano, G. (2020). Stability and change of basic personal values in early adolescence: A 2-year longitudinal study. *Journal of Personality*, *88*(3), 447–463. https://doi.org/10.1111/jopy.12502

White, K. M., Smith, J. R., Terry, D. J., Greenslade, J. H., & McKimmie, B. M. (2009). Social influence in the theory of planned behaviour: The role of descriptive, injunctive, and in-group norms. *British Journal of Social Psychology*, *48*(1), 135–158. https://doi.org/10.1348/014466608X295207

6 Climate Change Challenges in Australian Classrooms

Karina Rune, Kirsty A. O'Callaghan, Rubie Orman-Ditchfield, and Patrick D. Nunn

Positionality Statement

As professionals dedicated to addressing one of the most pressing global challenges of our time, we recognize the far-reaching implications of climate change for both present and future generations. Our backgrounds – Rune as a psychologist, O'Callaghan as a communication specialist/researcher, Orman-Ditchfield as a schoolteacher/researcher, and Nunn as a climate scientist – inform our collective perspective on the importance of education in equipping students with knowledge and skills to understand and tackle this complex issue. Teaching young people about climate change requires a delicate balance: aligning with mainstream scientific projections without causing undue anxiety. In this chapter, we draw on our diverse experiences in Australia to reflect on this educational dilemma and explore how it may unfold in the future.

Introduction

In recent decades, the issue of climate change in Australia has often been perceived as a problem affecting other regions, particularly poorer nations and neighbouring Pacific Island countries that are considered to be at the forefront of climate change impacts. This perception may have been influenced by the prevailing right-wing media in Australia, which downplayed the significance of climate change and its future implications for the country at a time when global awareness was increasing. While acknowledging that this dilemma is not unique to Australia, the country's geographical separation from most other industrialized nations has resulted in an evolution of climate change awareness and emphasis that is distinct from European and North American countries. Despite being similar in size to Europe and the contiguous United States, Australia's population is concentrated in its coastal regions, particularly in large cities. In fact, 90% of Australians occupy just 0.22% of the land area. Although the non-coastal areas of Australia are largely desert and sparsely populated, they play a crucial role in agricultural production for both domestic consumption and export.

DOI: 10.4324/9781003494416-7

Additionally, mining is an important contributor to the Australian economy, and most is located in remote parts of Australia.

Three distinct characteristics about Australia are worth noting. First, the high degree of urbanization of people in Australia means that, like urban populations elsewhere, many urban Australians consider themselves immune from climate change as they occupy well-managed environments, making them inherently sceptical about climate predictions. Second, agricultural productivity in non-urban Australia has long been influenced by climate variability, including multi-year droughts attributed to the El Niño – Southern Oscillation (ENSO), which makes discussions about the attribution of climate impacts contentious but has also brought discussions about climate change to the forefront of conversations about Australia's future. Third, the proportionately large contribution of mining revenues to the Australian economy has sparked debates about both domestic coal power and the ethical implications of exporting substances like high-emissions (dirty) coal that will exacerbate global climate change. The purpose of this chapter is to examine aspects of climate change in Australia, with a particular emphasis on the perspectives and experiences of young Australians. The term "young people" is socially constructed and subject to change as it is situated between childhood and adulthood. In this chapter, we use the term to refer to schoolchildren aged 12 to 18.

Context

Australia is currently grappling with increasing impacts of climate change, and the growing urgency of addressing this issue is becoming widely recognized. Yet climate change remains a politically divisive topic in Australia, with divergent views among political parties which have hindered the formulation and implementation of effective climate policies. The development of a high school climate change curriculum in Australia has evolved over time, reflecting a growing recognition of the importance of climate education. Yet despite efforts to centre climate change concepts within the curriculum, the extent and depth of education have varied widely.

Australia is known for its vast and diverse landscapes, ranging from deserts to rainforests, and its unique flora and fauna. Yet these ecosystems are increasingly under threat owing to the increasingly visible impacts of climate change with observable shifts in weather patterns, more frequent and intense heatwaves, prolonged droughts, altered rainfall distribution, and increased frequency of extreme weather events. These changes have far-reaching ramifications, affecting the environment, agriculture, water resources, coastal communities, and human health. The effects of climate change have been experienced unevenly with vulnerable groups including young people disproportionately affected (Hoffmann et al., 2019; Hughes, 2011; Rahman & Zafarullah, 2020).

The political landscape surrounding climate change in Australia is multifaceted and characterized by a diversity of perspectives and policy approaches. While some political leaders and parties advocate for ambitious emissions reduction targets and investment in renewable energy, others prioritize economic interests tied to an increased use of fossil fuels such as coal, oil, and natural gas. This divide has resulted in ongoing debates about the urgency of climate change action, the role of international agreements, and the precise nature of an approach to mitigation that addresses both environmental concerns and economic considerations (Baer, 2023; Beeson & McDonald, 2013; Hornsey et al., 2022).

In Australia, major political parties, such as the Liberal, National, and Labor parties, have often avoided undue public debate about climate change due to radically differing views within party rooms and among the Australian public. The discourses of Australian political leaders over the past two decades, particularly from the Liberal–National (conservative) coalition, have often favoured climate change denial and scepticism. This stance has led to disagreements about the urgency of addressing climate change and the continued growth of resource-intensive, high-consumption lifestyles (Ali et al., 2020; Eckersley, 2013). In 2021, Australia ranked last for climate action among 193 countries and received very low scores on the *Climate Change Performance Index* for its commitment to reducing greenhouse gas emissions, increasing renewable energy, and implementing sustainable climate policies (Puertas & Marti, 2021). As a consequence, more Australians are now voting for candidates, often unaligned, who support stronger climate action, indicating a weakening of conservative opposition to climate science (McAllister, 2023; Stevenson, 2022).

In addition to public sentiment, there is also growing interest among younger Australians in political activism and environmental advocacy. Following the devastating "Black Summer" bushfires of 2019–2020, youth-led climate action, inspired by Swedish activist Greta Thunberg, gained momentum in Australia with movements like the School Strike 4 Climate movement and Extinction Rebellion (Figure 6.1). Thousands of high school students participated in strikes to demand urgent action on climate change and holding their government accountable for inadequate environmental policies. This trend reflects increased youth involvement in climate action and the need for a shared understanding of the challenges posed by a changing climate (Jones & Lucas, 2023).

Climate change education initiatives have been implemented in Australian schools since the early 2000s, aiming to raise awareness and promote sustainable practices among young people. Yet mandated requirements for climate change education have been notably absent from policies and updates to the *National Curriculum* that all Australian high schools are required to teach (Dawson et al., 2022; Whitehouse & Gough, 2022). Both the inconsistency and the limited scope of climate change education reflect broader challenges in curriculum integration and professional development, which has led to calls for holistic educational approaches that transcend disciplinary boundaries (Beasy et al., 2023). During

Figure 6.1 School Strike for Climate Protesters, Sydney, Australia, March 15, 2019

2022–2023, the rollout of the new National Curriculum 9.0 marked an increase in climate change content across numerous subjects, providing more students with opportunities to learn about climate change while also giving teachers clearer direction on how and where to teach it. The new curriculum also presents climate change in innovative ways, integrating it into subjects like mathematics and incorporating the campaigns of youth activists into the study of global citizenship.

Yet the delivery of climate change education in Australian is not prescriptive and leaves the precise nature and amount of delivery largely to individual educators. This results in an uneven implementation influenced to some extent by educators' personal motivations and beliefs (Waldron et al., 2019), putting Australia even further behind those countries where it is recognized that climate change education should go beyond the dissemination of climate science and include the knowledge, skills, and attitudes necessary for adaptation (Reilly et al., 2024).

Young Australians and Climate Change

During the course of their lives, today's children and young people will bear a greater burden of climate change yet their voices and perspectives have commonly been overlooked in research and public discourse. Climate change affects young people in many ways. First, they are particularly sensitive to economic transformations, which can disrupt their activities, prospects, and ambitions.

Second, young people will experience more extreme climate change events than previous generations. Finally, young people bring knowledge, ideas, dynamism, and political activism to the table, making their participation crucial in addressing climate change challenges. Engaging a youth audience in climate change discussions can be challenging due to the complexities of climate science, limited education, prevailing uncertainty, anxiety, and media clutter. Nonetheless, young people demonstrate a capacity for both individual and systemic transformative change, and their engagement, resilience, and diverse approaches contribute to tackling climate change issues (Arnot et al., 2023; Barford et al., 2021; Henn et al., 2022; Hohenhaus et al., 2023).

Young people in Australia increasingly view climate change as a critical issue impacting their future (Chiw & Ling, 2019; Hickman et al., 2021). Climate change concern among young Australians is associated with psychological distress, negative future outlook, and often a profound dissatisfaction with the nature and pace of government action (Arnot et al., 2023; Teo et al., 2024). Furthermore, young Australians have begun to express the importance of climate literacy and a need for political consensus in achieving a just energy transition (Green, 2022). Specifically, young Australians identify challenges in navigating climate change resulting from disempowerment, limited resources, and a lack of adult support and empathy (Chavez et al., 2024; Jones & Lucas, 2023). Many young Australians feel their opinions on climate change are not taken seriously and that the Australian government does not sufficiently recognize climate change as a pressing issue and lack the commitment to addressing it effectively.

Expressions of Eco-anxiety Among Young Australians

Eco-anxiety is a complex psychological phenomenon that disproportionately affects young people (Coffey et al., 2021; Hickman et al., 2021; Searle & Gow, 2010). Eco-anxiety describes the persistent fear of environmental doom in the context of climate change marked by a sense of loss, helplessness, and anger. Anxiety itself is an evolutionary alarm mechanism that aids individuals to prepare, find solutions, and cope with potential risks. Pathologizing eco-anxiety overlooks its potential as a motivator for action-driven solutions (Bhullar et al., 2022; Verplanken et al., 2020), so it is important to view eco-anxiety as a complex emotional response to the ecological crisis rather than a diagnosable mental health disorder.

Young people, being at a critical developmental stage, are more susceptible to the long-term impacts of stress and anxiety on their mental health. A study of young Australians found that climate change has detrimental effects on their mental health, producing worry, eco-anxiety, stress, hopelessness, and feelings of powerlessness (Gunasiri et al., 2022). Young Australians experience intense negative emotions, including frustration, anxiety, sadness, and anger when thinking about climate change, with 73% of young Australians identifying climate change as a main impact on their health. The experience of eco-anxiety among young Australians is deeply intertwined with their observations of the visible and tangible consequences of climate change including storms, floods, biodiversity loss, diminished air and water resources, and the prospect of inheriting a degraded planet (Boyd et al., 2024).

Engaging in climate action can have positive impacts on young people's mental health and well-being, providing social inclusion, mutual support, and collective opportunities for active coping strategies (Ojala & Bengtsson, 2019). Youth-led organizations such as the *Australian Climate Coalition* and *School Strike 4 Climate* are at the forefront in advocating for a healthier, more just and sustainable future in Australia. Eco-anxiety has been linked to increased climate action and engagement with news and politics in late adolescence, helping young Australians manage their anxieties about the future and instilling a sense of hope, optimism, and determination (Sciberras & Fernando, 2022).

How Young Australians Access Information About Climate Change

Discussions about climate change within family networks and among friends both influence and predict engagement in pro-environmental behaviours in young people (Wallis & Loy, 2021). While young Australians rate friends as the most trusted individuals for having more positive climate-related discussions, conversations with older people can evoke negative emotions like betrayal, uncertainty, and worry (Jones & Lucas, 2023). Family can provide a valuable platform for climate change discussions and facilitate the exchange of environmental concern between parents and young people (Baker et al., 2021; Trott, 2020). Yet, some young people report unhelpful climate change discussions with their parents with only 55% of Australian children and young people feeling that their opinions and concerns about climate change mattered (Chiw & Ling, 2019).

Schools have the potential to support children's awareness of climate change and their emotional responses by modelling sustainable changes, inviting engaged speakers, and supporting student activism (Tattersall et al., 2022; White et al., 2022). Teachers also play an important role in helping young people cope with and communicate about climate change. For example, Jones and Lucas (2023) found that young Australians who felt listened to and perceived their teachers as equally concerned were more likely to engage in climate conversations. Young Australians actively seek platforms to express their understanding and concerns about climate change, participating in activism through strikes, legal cases, and demands for government action. Visual communication, both in live protests and on social media platforms, has become a powerful tool for promoting climate action (Catanzaro & Collin, 2021; Spitzer et al., 2024).

Outside of high school, social media platforms like TikTok are popular sources of climate change information, providing opportunities for young

Australians to follow environmental activists, organizations, and influencers who share news, facts, and calls to action related to climate change (Hilder & Collin, 2022). Yet concerns exist about the dissemination of misinformation and disinformation on such platforms that might conceivably exacerbate rather than reduce young people's eco-anxiety (Arnot et al., 2024; Campbell et al., 2023). Despite the prevalence of social media, in-person conversations remain the preferred mode of climate communication among young Australians (Salguero et al., 2024).

Case Study: Young People's Views on Climate Change in Australian Schools

Understanding young people's perceptions, experiences, and emotional responses to climate change is important for developing effective strategies to mitigate the effects of eco-anxiety and help underwrite sustainable futures for young Australians. With this in mind, we conducted semi-structured interviews with a sample of 19 Australian schoolchildren (aged 14–17) recruited from high schools on the Sunshine Coast in Queensland, Australia. Thematic analysis revealed four main themes relating to (1) personal experiences and perceptions; (2) emotional impact; (3) information sources and communication; and (4) action and future outlooks. Each is discussed separately in the following.

Theme 1: Personal Experiences and Perceptions

Theme 1 explores participants' personal experiences with severe weather events and their perceptions of climate change. Most participants reported personal experiences with severe weather events including cyclones, floods, landslides, bushfires, storms, and droughts. Some participants were directly affected by these events, such as being evacuated or trapped in flooded homes. The impact on daily life, including school closures and transportation, was also discussed. Participants believed that weather events are more frequent and severe than in the past. For example, Participant TA (17 years) said:

> I think it's getting more severe. I feel like when it rains it's like way heavier, but then we'll go for like massive dry stints or like no rain for months. And then it'll just absolutely pound down and you get flooded and it's just really inconsistent.

Participants attributed changes in weather events to climate change and global warming (e.g. "Global warming, I guess. Climate change," Participant MM, 15 years). Overall, young people in the study agreed that weather events are increasing in both frequency and severity due to climate change and global warming.

The shared personal experiences highlight the long-lasting effects that changing climate events can have on young people's health and communities.

Theme 2: Emotional Impact

Theme 2 captures the emotional experiences of participants and their concerns about what most perceived as insufficient climate change action. Participants experienced stress, fear, helplessness, and anxiety when thinking about climate change. For example, Participant EP (15 years) felt helpless when thinking about climate change, "I guess I get stressed sometimes because I don't really know what to do about it either and like I'm not informed enough to make a decision about whether it's all true or not." Similarly, Participant MM (15 years) acknowledged the emotional toll of climate change and explained that she coped by pushing the issue to the back of her mind to avoid being overwhelmed, "Just a lot of stress and like helplessness. That sort of frozen anxiety. I guess that's just like being stuck."

The urgency and magnitude of the problem contributed to a sense of hopelessness and negativity (e.g. "There's a lot. . . . so it's kind of overwhelming to think about," Participant CW, 16 years). This emotional response reflects the impact of climate change and the perceived lack of control over the issue felt by many of the participants (e.g. "I cannot control it, so it makes you feel a little bit, like, I can't do anything, like, helpless, I suppose," Participant WB, 17 years).

Some participants also expressed concern for the well-being of animals, the destruction of land, and loss of biodiversity. Participant JS (17 years) noted, "but then you see it on social media like the polar bears and the snow and stuff like that. And other animals and their homes being affected, it does really like hit close to home."

While a few participants expressed hope and optimism due to the efforts being made globally to combat climate change, this was tempered by frustration at the perceived reluctance of older generations to address the issue. For example, Participant KP (15 years) explained:

I feel kind of hopeless in a sense, but also hopeful because being such a young person, I have to sort of rely on a bunch of old people to make decisions that will greatly affect my future and the way I live.

Theme 3: Information Sources and Climate Communication

Theme 3 explores sources of climate change information and communication among young people. Participants revealed varying levels of engagement within their social circles and the broader community. The role of social media, friends, family, and school in shaping discussions and actions related to climate change was evident throughout.

Participants emphasized the need for more comprehensive education about climate change in schools. For example, Participant MM (15 years) mentioned that climate change was rarely talked about at school:

> Not very often, I would say. I guess it is sort of a topic that teachers try to avoid because it can cause conflict and stuff. So, when we looked at it in class it was very scientific, it was more like "What is climate change? What's actually happening?" rather than like "What is causing it and what can be done about it?"

Likewise, Participant CW (16 years) mentioned that the topic was often dismissed by teachers ("and you try to ask your teachers and they kind of just dismiss it"). Some teachers were reported as not believing in climate change which further hindered climate conversation at school. For instance, Participant KP (15 years) stated:

> but I have had one of my teachers say that he doesn't believe that climate change is a very big thing and I just kind of sat there thinking to myself "OK." But it's not necessarily a thing that is discussed very often at my school.

The importance of climate change knowledge and collective action emerged as an important finding. Participant MM (15 years) believed that learning more about climate change in school is important, "Yeah, I think that learning about it is definitely important, even though I find it hard. Yeah, I think that the more we learn about it, the more we can actually do something about it."

Participants relied on various information sources such as news, social media, and discussions with family and friends, and teachers. Social media platforms like TikTok and Instagram played a significant role, but many expressed scepticism due to misinformation (e.g. "I quite often get my information through different sources, like reputable sources. I Google online and then I look at a bunch of different websites and then I look at Google Scholar as well," Participant KP, 15 years).

Friends and parents emerged as a source of information. Some participants said they would talk to their friends about climate change, though many avoided the topic as it was distressing. For example, Participant MM (15 years) said:

> It's just so overwhelming that we don't talk about it, but we do sometimes and they're [friends] probably the people who I can kind of talk to the most about it because they have such similar feelings to me. So, it's like we're getting super stressed, but getting super stressed about it together.

Similarly, Participant TA (17 years) said:

> Depends on the friend group, but for the most part, yes. Disappointing, a lot of anger. I guess that, we're going to have to live in a world where I'm going to have to surf in a hazard suit eventually.

Many participants mentioned that their parents have different views or were indifferent. For example, Participant JS (17 years) mentioned, "No, not really. We might like quickly say something about if it's on the news, but it's not like a topic you have at like the dinner table or something like that." Conversely, Participant TA (17 years) reported regular climate conversations with parents, "Yeah, all the time. They're really against it and really proactive about it. Mum's very anti single use plastic and all those little things which definitely add up over time."

Theme 4: Action and Future Impact

Theme 4 explores climate action strategies and future impacts. Participants revealed varying levels of engagement and beliefs about future environmental changes, including impacts on the economy, habitable land, and biodiversity.

Participants were uncertain about their ability to make a difference, instead expressing the need for more climate action from governments, big corporations, and people in power. They felt frustration and disappointment with the lack of progress, despite small improvements and increased awareness. Participant MM (15 years) highlighted the need for urgent and meaningful responses:

> I guess the thing that I get stuck on the most is it's not stopping and even though there are little improvements, more awareness here and there, it's like overall the big things that need to happen, like big corporations and people in power, isn't happening, and it just feels like that's never going to happen. And so, what can you even do about it? It's just like, hopeless.

Some participants expressed a commitment to take action themselves, even considering a career (e.g. "I'm aiming to be a psychologist, but I might end up going into environmental studies to try and help with this, just climate change in general." Participant TA, 17 years). Conversely, a few indicated a lack of personal engagement in climate communication relying on others for information (e.g. "I don't particularly go searching for it." Participant AK, 18 years).

Participants emphasized the importance of collective efforts and larger-scale actions (e.g. "I'm really hoping that more action is taken . . . I don't actually see how that's going to work, if it will work." Participant TA, 17 years). Participants also expressed frustration with the lack of action, particularly regarding CO_2

emissions and fossil fuel use (e.g. "We've known about it since the 70s and nothing's been done . . . It's just, really disappointing." Participant TA, 17 years).

Participants anticipated future environmental changes impacting both human and animal life. For example, Participant KP (15 years) expressed concern about the climate worsening over time and its impact on endangered species:

> Like 20 years in the future, we may not have very cold climates anymore, depending on if this weather and the temperature keeps rising, if polar caps keep melting and the water levels keep rising which just limits even more space. Then some of my favourite animals are endangered species, which means I will never get to see them thrive in the wild.

Participants worried about the future their children and future generations will inherit, believing they will bear the brunt of climate change. Climate change was seen as impacting lifestyle choices, including where to live and work. Participant JS (17 years) expressed concern about liveable spaces in the future, "probably won't want to live like close to the ocean and stuff like that, cause it could be affected, but you probably don't want to live like close to the like bushland either where it's like really dry." Similarly, Participant MM (15 years) felt stressed about her future:

> Because like I just think about how I want to have a life and a future. And just the idea that the world is getting harder to live in and then maybe even impossible to live in one day is just you can't even imagine that. And so yeah, just very stressful.

In contrast, some were hopeful about finding solutions to address climate change, such as green energy. Participant TA (17 years) expressed:

> I feel like things could get better. But if we keep going down that road it's just going to obviously keep getting worse and animals are going to go extinct, and we're not going to know what that kind of stuff is any more.

Looking to the Future

Climate change significantly impacts the mental health of young Australians. The disruption of daily routines, property damage, and health risks associated with severe weather events contribute to feelings of helplessness and uncertainty about the future. Recognizing environmental changes and their potential impacts on ecosystems and biodiversity contributes to pervasive worry and sadness among young people. The need for accurate information and a better understanding of the causes and effects of climate change is essential for young people's mental well-being. The frustrations and disappointments regarding inaction and reluctance of older generations to address climate change reflect the frustration and disillusionment among young people who often feel unheard and disregarded in the decision-making processes concerning climate change. Despite these concerns, young Australians desire collective action to address climate change and hope for positive change, underscoring their resilience and wish for a better future than that which appears most likely.

Today, we can project future climate change with considerable accuracy; yet, those projections vary due to different socio-economic scenarios. Drastic emissions cuts could slow or even reverse the current rise of temperature by the year 2100. Yet in comparatively high per-capita emitting nations like Australia, action to reduce emissions is currently lagging well behind what is required for twenty-first-century temperature stabilization (Bataille et al., 2016; Lu et al., 2022). Throughout the twenty-first century, Australia is expected to become hotter with more heatwaves, leading to heat stress in animals and plants, affecting agriculture and food production, human health, and energy systems. Changes in rainfall patterns and increased intensity of tropical cyclones will create novel stressors for life and food production. Prolonged dryness combined with higher temperatures will result in more wildfires (Clarke et al., 2016; Nishant et al., 2022). Rising temperature will also cause sea levels to rise around Australia, disproportionately impacting infrastructure since most Australians live near the coast. This will likely necessitate the inland relocation of people and activities, presenting significant challenges for the nation (Zhang et al., 2017).

Most climate change impacts are unavoidable, although prompt action can alter their severity and duration. The Australian national science agency (CSIRO) provides reliable projections and localized scenarios to inform policymakers and local authorities. Those projections guide the development of policies aimed at reducing climate change impacts while protecting the economy. Responses to climate change include mitigation, which addresses the causes of climate change, and adaptation, which tackles unavoidable future impacts. Australian national policy is directed at mitigation, especially the phasing out of fossil fuels in favour of renewable energy sources. In contrast, adaptation policies are generally left to local authorities to implement, leading to varied responses across different regions.

To support young Australians to cope with the impacts of twenty-first-century climate change, we must provide accurate information, foster emotional resilience through mindfulness practices and open dialogue, encourage engagement in climate action, and offer safe spaces for young people to express their concerns and hopes. Moreover, equipping them with practical adaptation skills, connecting them with mentors, and promoting inclusive climate education can empower them to become active contributors to sustainable solutions. By implementing these strategies, educators, families, communities, and policymakers can help young Australians not only cope with the effects of climate change but also participate in creating a sustainable future. Additionally, peer-to-peer communication and intergenerational dialogue is gaining traction, where young people influence the attitudes, consumption patterns, and political engagement of their caregivers. Young people's sense of urgency and frustration reflect a global call for transformative climate action amid climate change uncertainties. Effective communication, educational strategies, and dedicated efforts to empower young Australians are essential components of a sustainable future.

References

Ali, S. H., Svobodova, K., Everingham, J.-A., & Altingoz, M. (2020). Climate policy paralysis in Australia: Energy security, energy poverty and jobs. *Energies, 13*(18), 4894. https://doi.org/10.3390/en13184894

Arnot, G., Pitt, H., McCarthy, S., Cordedda, C., Marko, S., & Thomas, S. L. (2024). Australian youth perspectives on the role of social media in climate action. *Australian and New Zealand Journal of Public Health, 48*(1), Article 100111. https://doi.org/10.1016/j.anzjph.2023.100111

Arnot, G., Thomas, S., Pitt, H., & Warner, E. (2023). "It shows we are serious": Young people in Australia discuss climate justice protests as a mechanism for climate change advocacy and action. *Australian and New Zealand Journal of Public Health, 47*(3), 100048. https://doi.org/10.1016/j.anzjph.2023.100048

Baer, H. A. (2023). Climate change and capitalism, climate dystopia, and radical climate futures. *The Journal of Australian Political Economy, 91*, 107–127.

Baker, C., Clayton, S., & Bragg, E. (2021). Educating for resilience: Parent and teacher perceptions of children's emotional needs in response to climate change. *Environmental Education Research, 27*(5), 687–705. https://doi.org/10.1080/13504622.2020.1828288

Barford, A., Coombe, R., & Proefke, R. (2021). Against the odds: Young people's high aspirations and societal contributions amid a decent work shortage. *Geoforum, 121*, 162–172. https://doi.org/10.1016/j.geoforum.2021.02.011

Bataille, C., Waisman, H., Colombier, M., Segafredo, L., Williams, J., & Jotzo, F. (2016). The need for national deep decarbonization pathways for effective climate policy. *Climate Policy, 16*, S7–S26. https://doi.org/10.1080/14693062.2016.1173005

Beasy, K., Jones, C., Kelly, R., Lucas, C., Mocatta, G., Pecl, G., & Yildiz, D. (2023). The burden of bad news: Educators' experiences of navigating climate change education. *Environmental Education Research, 29*(11), 1678–1691. https://doi.org/10.1080/13504622.2023.2238136

Beeson, M., & McDonald, M. (2013). The politics of climate change in Australia. *Australian Journal of Politics & History, 59*(3), 331–348. https://doi.org/10.1111/ajph.12019

Bhullar, N., Davis, M., Kumar, R., Nunn, P., & Rickwood, D. (2022). Climate anxiety does not need a diagnosis of a mental health disorder. *The Lancet: Planetary Health, 6*, e383. https://doi.org/10.1016/S2542-5196(22)00072-9

Boyd, C. P., Jamieson, J., Gibson, K., Duffy, M., Williamson, M., & Parr, H. (2024). Eco-anxiety among regional Australian youth with mental health problems: A qualitative study. *Early Intervention in Psychiatry*. https://doi.org/10.1111/eip.13549

Campbell, E., Uppalapati, S. S., Kotcher, J., & Maibach, E. (2023). Communication research to improve engagement with climate change and human health: A review. *Frontiers in Public Health, 10*, Article 1086858. https://doi.org/10.3389/fpubh.2022.1086858

Catanzaro, M., & Collin, P. (2021). Kids communicating climate change: Learning from the visual language of the SchoolStrike4Climate protests. *Educational Review, 75*(1), 9–32. https://doi.org/10.1080/00131911.2021.1925875

Chavez, K. M., Quinn, P., Gibbs, L., Block, K., Leppold, C., Stanley, J., & Vella-Brodrick, D. (2024). Growing up in Victoria, Australia, in the midst of the climate emergency. *International Journal of Behavioral Development, 48*(2), 125–131. https://doi.org/10.1177/01650254231205239

Chiw, A., & Ling, H. S. (2019). Young people of Australia and climate change: Perceptions and concerns. *Millennium Kids*, 1–31.

Clarke, H., Pitman, A. J., Kala, J., Carouge, C., Haverd, V., & Evans, J. P. (2016). An investigation of future fuel load and fire weather in Australia. *Climatic Change, 139*(3–4), 591–605. https://doi.org/10.1007/s10584-016-1808-9

Coffey, Y., Bhullar, N., Durkin, J., Islam, M. S., & Usher, K. (2021). Understanding eco-anxiety: A systematic scoping review of current literature and identified knowledge gaps. *The Journal of Climate Change and Health, 3*, 100047. https://doi.org/10.1016/j.joclim.2021.100047

Dawson, V., Eilam, E., Tolppanen, S., Assaraf, O. B., Gokpinar, T., Goldman, D., Putri, G., Subiantoro, A. W., White, P., & Quinton, H. W. (2022). A cross-country comparison of climate change in middle school science and geography curricula. *International Journal of Science Education, 44*(9), 1379–1398. https://doi.org/10.1080/09500693.2022.2078011

Eckersley, R. (2013). Poles apart?: The social construction of responsibility for climate change in Australia and Norway. *Australian Journal of Politics & History, 59*(3), 382–396. https://doi.org/10.1111/ajph.12022

Green, M. (2022). "We need to care about this, and yes the facts are terrifying": Understanding young people's perspectives about energy transition and climate adaptation in regional Australia. *Children, Youth and Environments, 32*(2), 125–144. https://doi/10.1353/cye.2022.0018

Gunasiri, H., Wang, Y. F., Watkins, E. M., Capetola, T., Henderson-Wilson, C., & Patrick, R. (2022). Hope, coping and eco-anxiety: Young people's mental health in a climate-impacted Australia. *International Journal of Environmental Research and Public Health, 19*(9), Article 5528. https://doi.org/10.3390/ijerph19095528

Henn, M., Sloam, J., & Nunes, A. (2022). Young cosmopolitans and environmental politics: How postmaterialist values inform and shape youth engagement in environmental politics. *Journal of Youth Studies, 25*(6), 709–729. https://doi.org/10.1080/13676261.2021.1994131

Hickman, C., Marks, E., Pihkala, P., Clayton, S., Lewandowski, R. E., Mayall, E. E., Wray, B., Mellor, C., & Van Susteren, L. (2021). Climate anxiety in children and young people and their beliefs about government responses to climate change: A global survey. *The Lancet Planetary Health, 5*(12), e863–e873. https://doi.org/10.1016/S2542-5196(21)00278-3

Hilder, C., & Collin, P. (2022). The role of youth-led activist organisations for contemporary climate activism: The case of the Australian Youth Climate Coalition. *Journal of Youth Studies, 25*(6), 793–811. https://doi.org/10.1080/13676261.2022.2054691

Hoffmann, A. A., Rymer, P. D., Byrne, M., Ruthrof, K. X., Whinam, J., McGeoch, M., Bergstrom, D. M., Guerin, G. R., Sparrow, B., Joseph, L., Hill, S. J., Andrew, N. R., Camac, J., Bell, N., Riegler, M., Gardner, J. L., & Williams, S. E. (2019). Impacts of recent climate change on terrestrial flora and fauna: Some emerging Australian examples. *Austral Ecology, 44*(1), 3–27. https://doi.org/10.1111/aec.12674

Hohenhaus, M., Boddy, J., Rutherford, S., Roiko, A., & Hennessey, N. (2023). Engaging young people in climate change action: A scoping review of sustainability programs. *Sustainability, 15*(5), Article 4259. https://doi.org/10.3390/su15054259

Hornsey, M. J., Chapman, C. M., & Humphrey, J. E. (2022). Climate skepticism decreases when the planet gets hotter and conservative support wanes. *Global Environmental Change-Human and Policy Dimensions, 74*, Article 102492. https://doi.org/10.1016/j.gloenvcha.2022.102492

Hughes, L. (2011). Climate change and Australia: Key vulnerable regions. *Regional Environmental Change, 11*, 189–195. https://doi.org/10.1007/s10113-010-0158-9

Jones, C. A., & Lucas, C. (2023). "Listen to me!": Young people's experiences of talking about emotional impacts of climate change. *Global Environmental Change-Human and Policy Dimensions, 83*, Article 102744. https://doi.org/10.1016/j.gloenvcha.2023.102744

Lu, J. B., Chen, H. L., & Cai, X. Y. (2022). From global to national scenarios: Exploring carbon emissions to 2050. *Energy Strategy Reviews, 41*, Article 100860. https://doi.org/10.1016/j.esr.2022.100860

McAllister, I. (2023). Party explanations for the 2022 Australian election result. *Australian Journal of Political Science, 58*(4), 309–325. https://doi.org/10.1080/10361146.2023.2257611

Nishant, N., Ji, F., Guo, Y. M., Herold, N., Green, D., Di Virgilio, G., Beyer, K., Riley, M. L., & Perkins-Kirkpatrick, S. (2022). Future population exposure to Australian heatwaves. *Environmental Research Letters, 17*(6), Article 064030. https://doi.org/10.1088/1748-9326/ac6dfa

Ojala, M., & Bengtsson, H. (2019). Young people's coping strategies concerning climate change: Relations to perceived communication with parents and friends and proenvironmental behavior. *Environment and Behavior, 51*(8), 907–935. https://doi.org/10.1177/0013916518763894

Puertas, R., & Marti, L. (2021). International ranking of climate change action: An analysis using the indicators from the climate change performance index. *Renewable and Sustainable Energy Reviews, 148*, 111316. https://doi.org/10.1016/j.rser.2021.111316

Rahman, R., & Zafarullah, H. (2020). Impact of climate change on human health: Adaptation challenges in Queensland, Australia. *Climate Research, 80*(1), 59–72. https://doi.org/10.3354/cr01591

Reilly, K., Dillon, B., Fahy, F., Phelan, D., Aarnio-Linnanvuori, E., De Vito, L., Gnecco, I., Gopinath, D., Holmes, A., Laggan, S., & Mansikka-aho, A. (2024). Learning from young people's experiences of climate change education. *Geography, 109*(1), 44–48. https://doi.org/10.1080/00167487.2024.2297616

Salguero, R. B., Bogueva, D., & Marinova, D. (2024). Australia's university Generation Z and its concerns about climate change. *Sustainable Earth Reviews, 7*(1), Article #8. https://doi.org/10.1186/s42055-024-00075-w

Sciberras, E., & Fernando, J. W. (2022). Climate change-related worry among Australian adolescents: An eight-year longitudinal study. *Child and Adolescent Mental Health, 27*(1), 22–29. https://doi.org/10.1111/camh.12521

Searle, K., & Gow, K. (2010). Do concerns about climate change lead to distress? *International Journal of Climate Change Strategies and Management, 2*(4), 362–379. https://doi.org/10.1108/17568691011089891

Spitzer, J., Grapsas, S., Poorthuis, A. M. G., & Thomaes, S. (2024). Supporting youth emotionally when communicating about climate change: A self-determination theory approach. *International Journal of Behavioral Development, 48*(2), 113–124. https://doi.org/10.1177/01650254231190919

Stevenson, R. B. (2022). Climate movements, learning and the politics of climate action in Australia. *Australian Journal of Adult Learning, 62*(3), 424–443. http://doi.org/10.3316/informit.833875694688881

Tattersall, A., Hinchliffe, J., & Yajman, V. (2022). School strike for climate are leading the way: How their people power strategies are generating distinctive pathways for leadership development. *Australian Journal of Environmental Education, 38*(1), 40–56. https://doi.org/10.1017/aee.2021.23

Teo, S. M., Gao, C. X., Brennan, N., Fava, N., Simmons, M. B., Baker, D., Zbukvic, I., Rickwood, D. J., Brown, E., Smith, C. L., Watson, A. E., Browne, V., Cotton, S.,

McGorry, P., Killackey, E., Freeburn, T., & Filia, K. M. (2024). Climate change concerns impact on young Australians' psychological distress and outlook for the future. *Journal of Environmental Psychology, 93*, Article 102209. https://doi.org/10.1016/j.jenvp.2023.102209

Trott, C. D. (2020). Children's constructive climate change engagement: Empowering awareness, agency, and action. *Environmental Education Research, 26*(4), 532–554. https://doi.org/10.1080/13504622.2019.1675594

Verplanken, B., Marks, E., & Dobromir, A. I. (2020). On the nature of eco-anxiety: How constructive or unconstructive is habitual worry about global warming? *Journal of Environmental Psychology, 72*, 101528. https://doi.org/10.1016/j.jenvp.2020.101528

Waldron, F., Ruane, B., Oberman, R., & Morris, S. (2019). Geographical process or global injustice? Contrasting educational perspectives on climate change. *Environmental Education Research, 25*(6), 895–911. https://doi.org/10.1080/13504622.2016.1255876

Wallis, H., & Loy, L. S. (2021). What drives pro-environmental activism of young people? A survey study on the fridays for future movement. *Journal of Environmental Psychology, 74*, Article 101581. https://doi.org/10.1016/j.jenvp.2021.101581

White, P. J., Ferguson, J. P., Smith, N. O., & Carre, H. O. (2022). School strikers enacting politics for climate justice: Daring to think differently about education. *Australian Journal of Environmental Education, 38*(1), 26–39. https://doi.org/10.1017/aee.2021.24

Whitehouse, H., & Gough, A. (2022). Gesturing not acting: Searching for policy guidance for Australian climate educators. *Australian Journal of Adult Learning, 62*(3), 376–394.

Zhang, X. B., Church, J. A., Monselesan, D., & McInnes, K. L. (2017). Sea level projections for the Australian region in the 21st century. *Geophysical Research Letters, 44*(16), 8481–8491. https://doi.org/10.1002/2017gl074176

7 Caring for Each Other in Uncertain Times

Teaching and Learning at the End of *A* World

*Dawn Wiseman, Terra Léger-Goodes,
Rebeca L. Esquivel, Emma Cognet,
Mitchell McLarnon, Catherine M. Herba,
Limin Jao, and Catherine Malboeuf-Hurtubise*

Positionality Statement

We are an interdisciplinary, multi-university team of researchers at different career stages in education and psychology who have come together to consider the question "How do we prepare for teaching and learning in the context of human-driven climate change?" We work together under the name AppROC-CHE (Applied Radical Optimism for Climate Change Hope in Education). It is a name chosen to reflect the bilingual (and frequently multilingual) reality of our setting in Québec. AppROCCHE is a homophone to *approche*, in English "approach." The name thus also reflects:

(1) the wicked nature of the problem with which we engage, where definite answers to climate change seem elusive but a multiplicity of approaches seem promising;
(2) the interdisciplinary nature of the inquiry we undertake, where we draw on both the more quantitative methods of psychology and the more qualitative methods of education to explore how different approaches might come together in fruitful ways with respect to the problem;
(3) the dynamic nature of how we come together, where we collectively work on AppROCCHE, but create space for other research and partnerships which return to, approach, and inform our collective efforts.

Coming together at the beginning of the SARS2-COVID19 pandemic forced us into care-full conversations before we could get to the "applied" part of our research. This time allowed us to consider the kind of work we want to commit to and clearly articulate our individual and collective accountability to young people, educators, ourselves, and our planet. We thus begin from the position that both education and psychology are grounded in an ethic of caring (Noddings,

DOI: 10.4324/9781003494416-8

1988) "that is directly concerned with the relations in which we all must live" (p. 219). Education and psychology are well positioned as partners for considering the dynamics at play as teachers and students grapple with uncertainties of climate change, and for offering ways of moving through the end of one world and into another (as yet) unknowable one through teaching and learning.

While "The End is Nigh" or "The End of *a* World" might seem hyperbolic in relation to what happens in classrooms, the words reflect a general sense of the state of a world where both multiple global crises are intersecting (UNDP, 2023) and the young people and educators with whom we work clearly express worry, anxiety, and despair in relation to a changing and uncertain climate. In fact, our collective work was inspired by a young person who, in response to a question about why their peers should care about climate change, replied: "Because it is going to kill us all" (personal communication, October 22, 2019). The example suggests that, as indicated by the name AppROCCHE, our work should focus on hope and action; considering with young people and their educators how we might imagine different solutions that lead to other possible worlds where the end is not nigh.

So, why would a research collective working on hope and action consider the end of *a* world?

First, because we believe in meeting young people and their educators where they are at. The sense of anxiety we hear from them does not arise in a vacuum. It is backed up by research and data they have access to through various sources (some less unreliable than others), including the media. For instance, in late 2023, two climate reports received significant attention in the news (see, e.g. Dance, 2023; Paddison, 2023; Thompson, 2023). One report, by the not-for-profit Climate Central (2023), indicated that the average global mean temperature as measured between November 2022 and October 2023 had increased 1.3°C over the pre-industrial baseline. The other, by the European Union's Copernicus Climate Change Service (2023), reported a slightly higher global mean increase of "1.46°C above the 1850–1900 pre-industrial average" (November 2023–Surface air temperature . . ., para. 4) between January and November 2023. Both indicated that 2023 was likely to be the hottest year on record. While not a height to aspire to, this record has subsequently been confirmed (Copernicus Climate Change Service, 2024; National Oceanic and Atmospheric Administration, 2024). Moreover, in a statement quoted in a number of pieces about the late 2023 reports, the Deputy Director of Copernicus, Samantha Burgess, noted that, when combined with data from the Intergovernmental Panel on Climate Change, 2023 was likely "the warmest year for the last 125,000 years" (e.g. Abnett & Dickie, 2023, para. 6). That's quite a headline. Whether the recent increased incidence and severity of forest fires, flooding, storms, etc., are ultimately shown to be directly caused by the rise in global mean temperature or only to correlate with it, the young people with whom we work can see, hear, and – in the case of 2023 wildfires in Canada (Voiland, 2023) – smell, taste, and feel (both physically and emotionally) that change. So can their teachers. They are experiencing, up close and personally, existential dread, and grief caused by the uncertainty of a world in the midst of a significant climate crisis (Cunsolo & Ellis, 2018; Doherty & Clayton, 2011; Gifford & Chen, 2017; Robbins & Moore, 2012). Their fears and anxiety are not unreasonable nor misplaced. Instead, they may be rational responses to not being able to know or predict what will happen next. In many ways, the world they know or have been taught about is ending, and the world that replaces it is not yet clear.

Second, we consider the end of *a* world, because we know the world is and has always been multiple, and that the impacts of global climate change and privilege are unevenly spread (Rouf & Wainwright, 2020; Wynter, 2003). For many people (non-human relatives and environments), the world has ended in countless ways and many times due to more immediate impacts of colonialism, capitalism, and Western imperial expansionism (Davis & Todd, 2017; Sultana, 2022b). As scholars in, and primarily from, the Global North, our current lack of preparation for teaching and learning in the context of human-driven climate change is largely related to the privilege experienced by people in Western, Educated, Industrialized, Rich, and Democratic (Henrich et al., 2010) countries, and as privileged people even within that context. The world we (as researchers and human beings) and the young people and educators with whom we work perceive as ending in relation to climate change is a world that has never existed for most people globally. We are only now experiencing the impacts of climate change directly because we have, until recently, been buffered by privilege from the violences of colonial processes.

Third, we consider the end of *a* world, because we believe the world will continue, even if we do not yet know what the new one will look like. We take seriously the calls from leading climate researchers (e.g. Gill, 2023; Hayhoe, 2023; Sultana, 2022a) and activists (e.g. Hayes & Kaba, 2023; Solnit & Young Lutunatabua, 2023) not to fall into despair but rather to focus on action, hope, and change. It is a radical approach (Lear, 2006) that recognizes the young people with whom we work, and their teachers require care as a "revolutionary vehicle" (Sultana, 2022a, p. 1) for moving towards action, hope, and change. Some scholars (e.g. de Oliveira, 2021) posit that for such change to occur, action might include "hospicing"; that is, extending care to a world that is ending, and our attachments to it, so that another, new one might arise. Care is key because it opens up space for possibilities, for collective imaginings, for exploring liminality between worlds, for disrupting boundaries and binaries, and for moving beyond success/failure narratives. So, while we may not know what a new world might look like or "what kind of world we are currently trying to understand" (Wiseman et al., 2024, p. 54) alongside young people in teaching and learning, we are certain there will be a new world to understand together.

So, while this chapter considers the end of *a* world as the context within which our inquiries exist, it is grounded in care as a means of responding to that context. Within the chapter, we consider how we have observed different ways in which ideas of care, caring, and being care-full exist in places where young people and their educators interact and think about climate change. In writing about this work, we begin with a consideration of care to focus on how caring relationships require the reciprocity of response and who/what is cared for. We present findings from two studies – one focused on teachers, the other on students – to show how each group is grappling with climate anxieties that can affect relationships of care, and we describe different ways care is enacted. We then suggest possible ways of navigating uncertainties that maintain/reestablish response and relationships (broadly conceived) so that teaching and learning might become places of care and hope for navigating towards a new world. Finally, we consider the longer-term implications of the work.

Care, the Ability to Respond, and Relationality

Care, the ethics of care, and caring are ideas taken up across disciplines, from philosophy to political science (see, e.g. Lévinas, 1998; Tronto, 1993). The concepts have a particular place in feminist thinking and theory, beginning with the unpaid labour of women and moving into inquiries about the relationships and tensions between the practical *caring for* and affective *caring about*, as well as dependencies and imbalances of power in caring relationships (Elliott, 2016; Tronto, 1993; Ungerson, 2006). Not surprisingly, care, the ethics of care, and caring are also taking up in fields related and/or perceived to focus on both practical and affective care, including the health sciences (e.g. Waldby, 2012), social work (e.g. Ellis et al., 2007), and education (e.g. Long, 2011). Academic work regarding care points to the concept "as a source of critical tension in . . . theory, policy and practice" (Rummery & Fine, 2012, p. 322). Thus, in discussing care, it seems important to outline the ideas of and conceptualizations of care we invoke in our own work. We recognize that such an exploration could take up several papers, and so note the following engagement with ideas of care is limited. We look forward to the conversations that emerge from it.

Nel Noddings (1984) provides an ethics of care that is grounded in the ontological and the fundamental nature of being in relationship. Within education, she positions care as a teacher's ability to respond to student needs in ways that reflect a deep empathy for what students are experiencing. In this relationship, there is someone who cares (usually the teacher) and someone who is cared for (usually the student). The onus for response lies more deeply with the teacher. Thus, there is some power imbalance in the caring, but both parties are necessary for caring to occur; that is, there is a need for some kind of mutuality of response, even if it is an uneven one.

In the work we do with young people and their educators, we have observed human impacts of climate change – eco-anxiety, eco-grief, and eco-paralysis (Cunsolo & Ellis, 2018; Doherty & Clayton, 2011; Gifford & Chen, 2017;

Klocker et al., 2021; Robbins & Moore, 2012) – acting as barriers to teaching and learning about/within the context of climate change. In other words, anxiety and other impacts can interrupt the caring response in the teacher–student relationship via reticence to engage with the difficult knowledge (Britzman, 2000) and existential threat of climate change. We see a need to tackle these barriers for the well-being of the people involved but also because teaching and learning are identified as key pillars in addressing climate change at local, national, and international levels (Arnold, 2020; Field et al., 2019, UNESCO, 2019). It is unlikely that teaching and learning can fulfill this key role without the kind of care Noddings (1984, 1988) suggests if teachers and students are unable to move with hope towards a new world (Ojala, 2017; Stein et al., 2020).

Our experiences with young people and their educators make us wonder if the relationships involved in considering climate change in teaching and learning move beyond the binary of teacher–student (no matter how caring it is). At a human level, this conception of caring relationships being between two people makes sense. At a curricular level, in terms of teaching and learning, it ignores the situational praxis Aoki (2005a) suggests can lead to curriculum as a field (Aoki, 2005b) emergent in relation to the world. In such a field, teachers and students can walk together care-fully and the space for exploring is sufficiently large that it is possible to negotiate or move around the kinds of barriers that can disrupt teaching and learning about climate change. That is, if one pathway through the field proves too anxiety-inducing, there are a multiplicity of other pathways that may still lead to understanding the field. In this kind of approach, the subject matter, field, and world are as present as teachers and students. They all exist in relationship to each other. They all respond to each other. In fact, in this kind of approach, "the world . . . needs our memory, our care, our intelligence, our work the 'continuity of [our] attention and devotion' . . . and understanding if it is to remain hale and whole" (Jardine et al., 2006, pp. 90–91).

From this perspective, we suggest that while Noddings' (1984, 1988) early notions of care – that focus exclusively on human relations – are helpful, there is a need to somewhat extend them. We recognize that in later work Noddings (2005) takes up a deeper relationality that extends caring to the other-than-human and what Western science would classify as inanimate. This shift opens up care as a response that "require[s] us to live moderately, sensitively, and responsibly" (Noddings, 2005, p. 138) with the planet and each other to assure our own survival, but still privileges the human. Noddings still does not conceptualize fully reciprocal relationships with non-human entities, or relationships that simultaneously involve the teacher, student, and the world as a whole that might lead to the kind of praxis intimated by Aoki (2005a, 2005b). So, while holding onto the notion of care, and the need for response, we also think with Papachese

Cree scholar Dwayne Donald's (2012) notion of ethical relationality where trusting, respectful, and caring relationships extend beyond the human to a responsive, multi-directional kinship that

> repeatedly acknowledge[s] and honour[s] the fact that the sun, the land, the wind, the water, the animals and the trees (just to name a few) are quite literally our relatives . . . [and we are] fully reliant on them for our survival.
>
> (p. 10)

In Donald's work, we find a deeper relationality and response that creates space for the kind of care-full hope that may support teachers and students in continued engagement in teaching and learning about and within the context of climate change. Donald's thinking emerges from a foundationally different philosophical orientation to the world, literally a different world view from that of Noddings. We suggest, as have others (e.g. Schuster et al., 2019), that a different world view may, in fact, be what is needed to address climate change and its broad impacts (but that exploration is beyond the scope of this chapter). Nonetheless, in the conception of relationality discussed by Donald, we find a world experiencing climate change in the sense of Aoki (2005a, 2005b) whereby a multiplicity of relationships exist to open up rather than shut down teaching and learning. We also find care as a "revolutionary vehicle" that Sultana (2022a) characterizes as

> the very stuff of life, whereby in the face of grief and desolation, we survive and carry on through generations because of transgressive love and mutuality that propels us forth, forcing us to take another breath, disallowing us to wither away, embracing us in the drowning over-whelm of distress and turbulence. Care and care ethics are thrown into sharper relief for survival, renewal, and regrowth. Living persists amidst dying, whether personal or planetary.
>
> (p. 1)

Sultana exemplifies how we, as a research team and human beings experiencing climate change, consider care as a means of moving to hope and action at the end of *a* world. We note that not only the kind of care and relationality described earlier is aspirational but also what we have heard described by teachers and learners. So, we work towards it based on what young people and their educators tell us about their experiences of struggling with climate change in places of learning: schools, gardens, forests, public programmes, etc. It also provides a means of considering the interdisciplinarity of the work that we do collectively and offers alternatives to the limits of care as conceived within particular disciplinary approaches.

Our Inquiries

In this section, we briefly share three lines of inquiry that we have engaged in over the last few years. The first two outline how care, caring, and response (or lack thereof) manifest for students and teachers. The final piece outlines some promising practices we have put in place in various contexts that attempt to open up to care-full relationality that extends beyond the human.

The first line of inquiry focuses on how teachers are navigating the uncertainty of teaching climate change. This qualitative inquiry builds from a 2021 assessment of teacher needs for engaging in teaching and learning in the context of human-driven climate change (McLarnon et al., In Press). Interviews dig deeper into how teachers deal with the uncertainty of climate change in formal and informal contexts (elementary through post-secondary) where teaching and learning about climate change occur. This section emerges from four interviews with educators primarily in K-12 contexts, one of whom is now a consultant for provincial ministry of education and one who works at the post-secondary level. Interviews lasted for approximately one hour, were transcribed, returned to the participants, and approved for inclusion in research.

The second line of inquiry explores how children are making sense of the uncertainty of climate change. This descriptive, qualitative inquiry consists of 12 interviews with children aged 8 to 12 (Léger-Goodes et al., 2023). Interviews took place in 2021 in relation to the question: What are the lived experiences of children in the context of climate change? The study describes both emotional experiences and coping strategies. Here, we focus on the findings of the interviews and discuss what these might indicate for potential interventions.

The third line of inquiry presents some promising practices and interventions that centre care and relationality that we have and are taking up in different learning contexts.

Teachers

In our interviews, teachers reaffirmed that teaching and learning in the context of an ending world require many expressions of care. These include expressions of care that align with Nodding's (1998) notions of care, such as caring for and about themselves and their students, creating hope in spite of, and sometimes as a result of, fear, and attempting to inspire caring feelings and actions in students for their human and non-human communities. Our conversations with teachers provide insight into the challenges and opportunities that bring about a need for care in learning settings and demonstrate the different ways that caring happens in these spaces. The interviews surfaced instances of more uneven relationships of care where teachers take on the seemingly non-reciprocal role of carers; some instances of a more balanced caring exchange; the need for, and importance of caring networks; and finally, teachers' hopes to inspire students to care for the human and other-than-human.

Although findings suggest no uniformity to conceptions of care in climate change education among participants, all teachers we had conversations with discussed engaging in caring relationships from the beginning of planning to the implementation of learning interventions, and in the intentional conversations they have with students. In all instances, the teachers spoke of care-full selection of resources that resonated with/and challenged the values and beliefs of the lived experiences of their students. From elementary through higher education spaces, resources were selected with consideration to students' ages, curricular requirements, and the socio-economic realities of students and their families. The resources showcase a variety of experiences, through the use of short stories, poems, disciplinary texts, and materials developed by non-profit organizations. For instance, one teacher working in a polarized and climate sceptical high school community expressed taking extra care in utilizing resources and engaging in conversations that would not further disengage students from considerations of climate change. She remarked

> It's a difficult conversation to have with students in the context where I teach, because most of our students' parents would be employed by the oil and gas industry, or at least tangential to the oil and gas industry. So, our conversations tend to be less explicit, like [instead of] doing a climate change discussion more specifically, like weaving it through.
>
> (February, 2023)

In addition, teachers in both elementary and higher education spaces reported choosing course resources and creating lessons with care to give students the best understanding of the climate crisis and promote hope. A higher education educator summarized her hopes

> I would hope they see themselves in the web of actors and appreciate where they were and why they matter in a galvanizing way . . . and if we're asking students to go there, and we are giving them the climate crisis, there has to be something that comes as a part of that that says "you belong in this conversation, this is for you. And this is what makes what you know and what you do relevant to this bigger problem."
>
> (May, 2023)

This teacher's hope for her students is perhaps a step in the direction of the ethical relationality that Donald (2012) calls for. There is no question that teachers are engaging in care in these actions, but we see relation imbalance in the examples provided where course materials are identified and chosen exclusively by the teacher. While the sample size of teachers is small, as a group they are engaged and innovative educators who push at the edges of what is prescribed in official curricula and programmes of learning. Thus, their singular authority in choosing course materials is surprising given the interest young people have in climate change and the influence young people (such as Greta Thunberg and Autumn Pelletier) have played in bringing climate change discourse into the public sphere.

Most of the teachers also addressed becoming aware and working through their own feelings and concerns in teaching the climate crisis. Teachers in elementary and secondary contexts worked through their own fears about teaching a topic that is still considered controversial in some spaces. "At first I was scared of what parents would say, and that people would think I'm pushing an agenda . . . now, I don't fear showing my honest emotions to students. I am letting my anger fuel me" (February, 2023). Another teacher uses self-reflection to ensure she is caring for her students' as they enact climate actions. She stated about taking up actions,

> I don't have to be a perfect person when I'm performing actions to try and help with climate change . . . and I think I let that guide my teaching, trying to say to students that they're not going to be perfect, and that's okay.
>
> (February, 2023)

This exploration of their own emotional experiences of climate change is a form of self-care, pushing teachers to continue taking up teaching in the context of the world in authentic and invigorated ways. It may also signal a kind of reciprocal care in response to field, because the teachers feel they must talk about it, and in the talking find some of their fears relieved.

Only one of four participants spoke explicitly about engaging in reciprocal care with her students. This higher education teacher cited an example in which, overcome in a moment of collective grief for the end of the world, teacher and students cried together. In contrast, the remaining elementary and secondary school teachers, perhaps a result of the inherent power imbalance brought about by the age difference between teacher and learner, reported experiences of care primarily through the lens of carers. Despite this lens, these teachers spoke of the hope they gained in the face of their students' own hope regarding the climate crisis. One elementary teacher stated

> I feel grateful to be teaching elementary school because they still have a little glimmer in their eyes. I think I'm okay now, because I'm working with little people . . . they're so cute. And they have a little sparkle [in the face of] climate change.

We suggest that this hope may itself be evidence of reciprocal care, as teachers draw comfort and inspiration from their students.

Finally, findings suggest teachers are also looking for reciprocal care outside of teacher–student relationships, both at an institutional level and through communities of practice. One elementary school teacher stated "I need to know someone's got my back and will amplify my voice. I need a community of open-minded individuals who will share values around this." Half of the interviewees felt they had access to academic and emotional support from their employers, communities of practice, or teaching teams. The other two teachers felt that support for educators was scarce.

As teachers spoke to us of their experiences in teaching about climate change, all the different ways in which caring and care-full practices are taken up surfaced. These expressions of care focused primarily on supporting intentional conversations between students and teacher, providing students with responsive and hopeful learning resources, and mitigating difficult experiences of learning in the face of the end of *a* world. Despite no uniformity in conceptions of care among the teachers, they centred primarily around human relations and positioned the teacher as carer and student as the one cared for. Overall, findings from these interviews suggest that care is integral to all aspects of teaching for climate change for both the teacher and the student, and teachers need to participate in reciprocal caring relationships to remain hopeful and engaged in their teaching practices.

Children

The present findings have been previously published but will be presented in relation to the notions of care discussed in this chapter, as this was not done in the initial publication. The results are those of 15 online semi-structured interviews with children between the ages of 8 and 12 from Quebec, done in 2022 (Léger-Goodes et al., 2023). Through the thematic analysis of the interviews using the software QDA Miner, it emerged that these children were aware of climate change impacts, learning about them from school and media sources. They mostly understood the concrete consequences of climate change, the impacts of pollution on ecosystems, from the age of 8 years old. The children interviewed experienced a wide range of emotions emerging from this awareness. Most expressed sadness for animals and people who experience the direct effects of climate change, anger towards people who pollute and past generations, as well as fear and anxiety for their future. This care for animals was particularly present in the interviews, as this 10-year-old child told us:

> It's a little bit, just like thinking what's going to happen, and if it's nothing, if the whole like the . . . Well, the ice melts, and then there's no polar bear. No . . . Most of the animals are dead . . . So sometimes it is kind of . . . well . . . Scary.

In these emotions, we can see how their sense of care extends to relations beyond the human and the current moment in time. Such relationality and care were also evident in shared feelings of hope in humanity, gratitude for not experiencing consequences firsthand, and a sense of connection with other children who are also preoccupied with climate change. For example, one 10-year-old participant told us:

> Well, I also feel grateful that I can . . . That I'm not in one of those places where it really is a big problem because it's really hard when that happens, and I'm grateful that . . . Well, I'm really lucky that I still survive and that it's kinda still easy for us, for my family.

Furthermore, the thematic analysis of interviews identified various coping strategies employed by the children. First, some coped with their emotions by searching for information about and acting to mitigate the effects of climate change (problem-focused coping):

> Well, most of the time, well, either I'm more careful, of course, but actually, to feel better, like sometimes after school, I go walk around, . . . and I see garbage everywhere, like pieces of plastic . . . and I find that disgusting. . . . And we just like pick up what we see and put it in the garbage.
>
> (participant, 11 years old)

Second, some tried to reduce their emotions by avoiding the subject, diminishing the severity of the threat of climate change: "[I would tell a friend] Not to worry, we still have a lot of time to live" (participant, 10 years old) or by seeking reassurance from their parents or friends (emotion-focused coping). Finally, a few also tried to re-frame the problem, making it more accessible to them, showing hope in humanity and technology while recognizing the threat (meaning-focused coping):

> Yeah, I, I'm hopeful that in a few decades it will be better . . . that everything will be . . . well not . . . there will have been ravages too, but that now it will start to reconstruct. Climate changes will reduce a bit.
>
> (participant, 11 years old)

Caring about climate change in the absence of a reciprocal relationship with an adult might lead children to avoid engaging with these emotions and the subject. It is difficult to hear children say things like:

> I'm angry at previous generation because actually, it's us, it's us who have to live with this . . . Like people who are 70 will die in like 30 or 20 years even, you know . . . it's me who is 11 years old that will live in 40 years, 50 years, 60 years . . . Probably I'm going to live that again and again because others who are 70 years old today will have passed away when I'll be 70 years old . . . But I will be there, I will see everything that degrades.
>
> (participant, 11 years old)

However, expressing care instead of dismissing these experiences, while promoting empowerment might allow children to hope for another world. One way to express care is to validate the anger or despair children are feeling with regard to the future, which may, once explored, lead to finding meaning and hope once again.

Finally, findings from the interviews suggest young people's needs for responsive adults and safe spaces to express their emotions about climate change. When exposed to difficult information and emotions about climate change, one participant told us: "I just talk about it to someone close and they say that yes it's horrible, and I don't know why it really relieves me to talk about it like that." In a different manner, responding to the same question, one participant told us:

> Well, I felt really sad, but I don't remember doing something. It was just really, really sad. [Yeah, you just really felt sad. Did you let it out and talk about it or cry a little?] No, I think I kept it in.

Being able to express these emotions and feel cared for and validated by parents or teachers could enable children to remain engaged with the issue of climate change. Children may also benefit from opportunities to take collective action regarding climate change so that their voices will be heard and responded to. These instances strongly suggest the need for care, response, and relationality to support their effective and affective negotiation of climate change.

The interviews underline that children should be getting the message – from educators, from parents, and from each other – that their emotions are valid and normal within the context of climate change. However, it is important that such messages are caring and do not lead young people to become paralysed, overwhelmed, or develop further mental health problems. As such, focusing on emotional expression and validation, empowerment, access to age-appropriate information, and caring for the next generations are central to our approaches. With adequate tools, children and adults can work together to transition towards a new world: "I think it might get worse or better, but then people will start to act and it might, well, get really better. I hope so" (participant, 10 years old).

Promising Practices

Ojala (2012) has identified that individual action can overburden children and is associated with depression and anxiety symptoms. And while avoidance may be adaptative in small doses (Pihkala, 2022), overly relying on this method can be associated to less action and anxious symptoms in the long term (Ojala, 2012). When the situation is rather uncontrollable, care-full practices that create meaning and maintain hope are the most adaptative responses for children. Such approaches maintain their engagement, feelings of empowerment, and reported life satisfaction (Ojala, 2012). As such, it is suggested that educators consider caring teaching and learning practices that support the creation of meaning, constructive hope, collective sharing, engagement, and self-efficacy. Both individually and collectively, we have explored these kinds of practices in educational settings with the aim of addressing the anxiety that interrupts response in teaching and learning and thus care and relationality between teachers and students. Here, we briefly describe a number of promising practices that not only focus on young people but may also enact care for their educators. The practices arise from a number of projects, some of them not directly within AppROCCHE, but tangential and informing to our collective work. To clarify which of the authors have engaged directly with each project to date, we include names beside the subsections.

Philosophical Inquiry and Existential Psychology (Malboeuf-Hurtubise, Léger-Goodes, Herba)

Philosophical inquiry in schools promotes critical thinking, caring, and creative reasoning in children, which can promote meaning making. Initially used as an approach to teaching and learning, these philosophical inquiries involve classroom discussions around a certain question or theme (e.g. responsibility, what is nature?). Dialogues between students are encouraged to explore values and different points of view. Children can choose their own questions and bring the discussion in the direction they want. Although it was not initially developed with the aim of having an impact on children's mental health, our team's work has indicated that philosophical inquiry, when used as a proxy for existential therapy with children, improves their mental health (Malboeuf-Hurtubise et al., 2021a, 2021b). By fostering resilience and self-determination in children, the use of philosophy with children may promote greater engagement about climate change and perhaps reduce distressing levels of eco-anxiety for some of them. In turn, greater self-determination has been shown to lead to improved well-being in children and adults. Philosophical inquiry also helps children to reflect on existential questions such as those generated by climate change (Malboeuf-Hurtubise et al., 2024). Since the experience of eco-anxiety is deeply rooted in existential threats to one's identity, happiness, meaning, death, freedom, and isolation (Passmore et al., 2022), it can be important for children to discuss, explore, and think critically about these topics, while being adequately supported by caring adults, such as teachers and parents. Finally, our team's previous work brings us to conclude that when children are faced with existential questions, supporting them in reflecting and developing critical thinking skills about these questions improves their well-being and self-determination (Malboeuf-Hurtubise et al., 2021a, 2021b).

Art-Based Interventions (Malboeuf-Hurtubise, Léger-Goodes, Herba)

With children, previous research has shown that art-based interventions facilitate expression, discussion, and awareness of one's emotions (Greenberg &

Harris, 2012), which is positively correlated with self-determination and mental health (Freilich & Shechtman, 2010). Art therapy with children also helps them gain a sense of control in situations where they typically have little or none (Beebe et al., 2010; Favara-Scacco et al., 2001). Given that the perception of lack of control is central to the experience of climate change and eco-anxiety, one can stipulate that using art to discuss and explore climate change and eco-anxiety could have similar effects on children by helping them to gain a sense of control and understanding over their emotions (Marks et al., 2023). Finally, a recent scoping review of interventions to address eco-anxiety identified creative expression as one of the main themes with the potential impact to help individuals promote a sense of community, which implies heightened caring (Baudon & Jachens, 2021).

Photovoice (Léger-Goodes, Malboeuf-Hurtubise, Herba, McLarnon)

While photovoice (Wang & Burris, 1997) is typically employed as a method for grassroots policy making and action, it is also considered a form of art therapy (Greene et al., 2016), and as a therapeutic approach that encourages civic responsibility, citizen action, engagement, and global social changes (Budig et al., 2018). In this way, photovoice encourages individual expression and self-determination through picture taking and analysis. As children are asked to discuss their respective pictures as a group, this approach promotes collective dialogue and discussions of community issues, which encourages caring and understanding among participants. With youth, photovoice has been used to foster engagement and citizen participation, the development of critical thinking, and moral judgement as well as social identity (Strack et al., 2004).

In recent years, photovoice has been used in research with the specific aims of discussing climate change and eco-anxiety, in youth and adults alike (Adams & Nyantakyi-Frimpong, 2021; Bulla & Steelman, 2016; Derr & Simons, 2020). With youth specifically, photovoice has also been useful to engage them, in a developmentally appropriate way, in discussions about the environment, sustainability, and the conservation of nature in their community (Catalani & Minkler, 2010). A recent study has also shown how photovoice can be used with children to help them learn about and understand climate change, with group discussions being conducive to expressing concern for how climate change has affected people around them such as family members or of their community (Lam & Trott, 2022).

Our team's preliminary work combining the arts, photovoice, and philosophical inquiry has shown that the climate crisis is an important topic to children and that they are consequently pleased when provided an emotional and physical space to explore how they feel, namely through artistic creation and dialogue. In a pre-pilot, we conducted with fourth-grade elementary school students, semi-structured interviews with students and their teacher indeed showed that combining the arts and philosophical enquiry lead to deeper conversations on climate change (Léger-Goodes et al., 2024). Photovoice was notably appreciated by the children, who were enthused at the idea of taking pictures, showing them to the group and reacting to those of their peers. Themes that were explored in the intervention allowed children and their teacher to explore different emotions related to the climate crisis: sadness, anger, fear, and stress. One student felt indifferent towards climate change but enjoyed taking part in artistic creation and group discussions on the topic. The teacher expressed how engaging students in creative activities and group discussions helped foster competence and self-determination, as students were free to create and discuss topics related to the intervention. The use of varied artistic mediums, such as drawing, sculpting, and photovoice, was appreciated by students and their teachers. In essence, while art therapy fostered emotional expression in the study participants, philosophical inquiry encouraged critical thinking and cognitive flexibility, highlighting the complementarity across practices.

Student-Directed Inquiry (Wiseman and Jao)

As demonstrated by educator interviews summarized earlier, the teaching and learning situations presented in learning spaces are most often care-fully defined and designed by adults, be they curriculum developers, resource creators, or teachers interpreting local curricular documents/programmes of study. And yet, young people engage in the world daily, arriving in these spaces with their own questions, their own ideas, their own interests, and their own caring relationships with the world. In work we have been doing in Grade 7 science and technology since 2017 with educators and students in a small, urban, private school, we extend care and response to students by following their lead in teaching and learning (Wiseman et al., 2020). We name what we do "student-directed inquiry" for two reasons. First, because it situates the project in relation to and in response to the extant literature. Here, research shows that student-directed approaches improve conceptual understandings, lead to better retention of understandings, and support competency development in multiple areas such as data collection and scientific reasoning (Bunterm et al., 2014; Hmelo-Silver et al., 2007; Minner et al., 2010). Second, because beyond the initial class where we collectively think about what inquiry might be, all other teaching and learning are directed by the students. Within this project, we have reproduced the positive outcomes of previous research and also demonstrated that by engaging in learning they direct students to find and engage in community that sometimes extends beyond their own classroom and school (Wiseman et al., 2020). In work we have presented at conferences (Driscoll et al., 2023; Rubin et al., 2022; Wiseman et al., 2022a, 2022b), we have shown how the inquiry while directed by students maps on to local required curriculum in relation to science and technology and all other disciplinary areas. For us, this finding indicates students care about the world in all its complexity. It also indicates they are cared for (or responded to) in teaching and learning in ways that open up to the kinds of fertile learning fields proposed by Aoki (2005a, 2005b), and Jardine et al. (2006) as well as Donald's (2012, 2016) conception of relationality.

What we have come to understand as missing in the descriptor "student-directed inquiry" is the care and response required by teachers in such contexts. In our project, nothing proceeds or is dismissed without student consent. Teachers must listen deeply and be able to respond, in the moment and over the course of the learning that students want to do. Specifically, we provide multiple ways for students to tell us where they are at: questions in class, exit tickets to collect students' thoughts and ideas for next steps at the end of each class, small and large group conversations, co-creation of choice and strategies, etc. We have recently reached the point where students generate their own questions for defining what comes next.

Provides means for improved teaching and learning in general. What it also provides is care as a "radical vehicle" (Sultana, 2022a) which allows teachers and students to collectively explore difficult knowledge (Britzman, 2000) such as that presented by climate change. Within the project, students tend to propose a number of possible projects, and then choose one to work on together. In Year 1, they chose to explore the crisis of Missing and Murdered Indigenous Women (Jao et al., 2019; Jao et al., 2023; Wiseman & Jao, 2019). Over the years they have proposed topics such as societal inequity, women's health, racism, sexism, screen addiction, and (every year) climate change. While they still have not chosen climate change as the guiding topic for their annual exploration, it is clear that by providing a space in school where they are able to care for the world as it is, and they are cared for in return, we are supporting them in the ability to be hopeful about longer-term societal change (Ojala, 2017).

Implications

The uncertainty of the world we find ourselves in does not lend itself to firm conclusions. Like the young people and teachers with whom we work, we find ourselves caught up in the global impacts of climate change. Some days it seems too much and that the end may indeed be nigh. When we move through the despair and anxiety, it is clear that the research we do is about hope and action, and it is also one way we collectively engage in care for and caring about the world, each other, and teachers and students as we all try to make sense of the end of *a* world. In fact, we find our lack of clear ending somewhat hopeful; it underlines the ongoingness of our research and relationships (and the world), and forces us to think about the next steps, implications of, and possible cautions to our work.

The next steps for us are clear. Based on preliminary data from conversations with teachers and students, there is more work to do on what care-full, caring, and responsive teaching and learning with respect to climate change looks like.

While we have identified and described some promising practices to follow up on, the different ways these practices support care and response in teaching and learning require more exploration. Tools and resources for communicating these practices to teachers and learners are also required.

Implications are broad, but we focus here solely within the confines of this chapter. There is a general need for a shift in how climate change is taken up in teaching and learning. We suggest that shift should emphasize the kind of aspirational care and response we have described. Here, we imagine young people, their teachers, and the world coming together to form rich fields where emotions and anxiety are cared for within teaching and learning so that they are not barriers to teaching and learning but perhaps drivers for it. These rich and robust imagined fields of teaching and learning provide space for care and response as praxis, hope, and action. Young people will live with the unfolding realities of the climate crisis throughout their lives. It will be a clear relationship they will need to navigate, for better or worse. We thus propose that they lead the exploration of teaching and learning in these fields, and the movement from the end of *a* world, to the emergence of a new one.

While we have focused on the promising, we recognize that there is some need to be cautious about care. As Liboiron (2021) notes, "Care is not inherently good. It is an uneven relation and can contribute to and/or mitigate unevenness" (p. 115). In our conversations with teachers, despite the care we see in their practices, we wonder if in caring there is the possibility of teachers who may be extracting hope from young people, instead of dealing more directly with their own anxieties and emotions about existential threat. In addition, some of the conversations we had with teachers raise the question of whether it is possible to care too much; that in trying to carefully choose the resources and materials young people engage with in teaching and learning about climate change, they do not get a full exposure to severity of the crisis or are not permitted space to consider their roles as agents of change in the crisis. Here again, we underline the need for expanding care to Donald's (2012) concept of relationality, so that teachers and young people face each other as human beings facing the same existential threat and figure out how to explore fields together.

This last thought leads to perhaps the most solid thing we can suggest from our work: that travelling together towards a new world through teaching and learning is not so much about what we do, the choice of materials, or promising practices; but rather much more about how we can come to be together at the end of *a* world. In this sense, it is about the journey towards the new than the destination itself.

Acknowledgments

The work presented in this chapter has been supported by a number of grants. We thank the following agencies and organizations involved for their ongoing support of our research, and their willingness to allow us time to figure out how to navigate the challenges of interdisciplinarity.

- Bishop's University, Interdisciplinary Team Grants
- Fonds de recherche du Québec Société et Culture
- McGill University, Innovation Education Project Grants
- Social Science and Humanities Research Council of Canada, Partnership Engage Grants

References

Abnett, K., & Dickie, G. (2023, November 8). 2023 'virtually certain' to be warmest in 125,000 years – EU scientists. *RNZ News*. https://www.rnz.co.nz/news

Adams, E. A., & Nyantakyi-Frimpong, H. (2021). Stressed, anxious, and sick from the floods: A photovoice study of climate extremes, differentiated vulnerabilities, and health in Old Fadama, Accra, Ghana. *Health Place, 67*, Article 102500.

Aoki, T. T. (2005a). Curriculum implementation as instrumental action and as situational praxis (1983). In W. F. Pinar & R. L. Irwin (Eds.), *Curriculum in a new key: The collected works of Ted. T. Aoki* (pp. 111–123). Lawrence Erlbaum Associates.

Aoki, T. T. (2005b). Legitimating lived curriculum: Toward a curricular landscape of multiplicity (1993). In W. F. Pinar & R. L. Irwin (Eds.), *Curriculum in a new key: The collected works of Ted. T. Aoki* (pp. 199–215). Lawrence Erlbaum Associates.

Arnold, J. (2020, January 14). Québec government to abolish ethics and religions course. *The Canadian Jewish News*. www.cjnews.com/news/canada/quebec-government-to-abolish-ethics-and-religions-course

Baudon, P., & Jachens, L. (2021). A scoping review of interventions for the treatment of eco-anxiety. *International Journal of Environmental Research and Public Health, 18*(18).

Beebe, A., Gelfand, E. W., & Bender, B. (2010). A randomized trial to test the effectiveness of art therapy for children with asthma. *Journal of Allergy and Clinical Immunology, 126*(2), 263–266.

Britzman, D. P. (2000). If the story cannot end: Deferred action, ambivalence and difficult knowledge. In R. I. Simon, S. Rosenberg & C. Epppert (Eds.), *Between hope and despair: Pedagogy and the remembrance of historical trauma* (pp. 27–56). Rowan & Littlefield.

Budig, K., Diez, J., Conde, P., Sastre, M., Hernán, M., & Franco, M. (2018). Photovoice and empowerment: Evaluating the transformative potential of a participatory action research project. *BMC Public Health, 18*. https://doi.org/10.1186/s12889-018-5335-7

Bulla, B., & Steelman, T. (2016). Farming through change: Using photovoice to explore climate change on small family farms. *Agroecology and Sustainable Food Systems, 40*(10), 1106–1132. https://doi.org/10.1080/21683565.2016.1225623

Bunterm, T., Lee, K., Ng Lan Kong, J., Srikoon, S., Vangpoomyai, P., Rattanavongsa, J., & Rachahoon, G. (2014). Do different levels of inquiry lead to different learning outcomes? A comparison between guided and structured inquiry. *International Journal of Science Education, 36*(12), 1937–1959. https://doi.org/10.1080/09500693.2014.886347

Catalani, C., & Minkler, M. (2010). Photovoice: A review of the literature in health and public health. *Health Education & Behavior, 37*(3), 424–451.

Climate Central. (2023, November 9). *Earth's hottest we-month streak*. www.climatecentral.org/climate-matters/earths-hottest-12-month-streak-2023

Copernicus Climate Change Service. (2023, December 6). *Copernicus: November 2023 – Remarkable year continues with the warmest boreal autumn. 2023 will be the warmest year on record.* https://climate.copernicus.eu/copernicus-november-2023-remarkable-year-continues-warmest-boreal-autumn-2023-will-be-warmest-year#:~:text=2023%20will%20be%20the%20warmest%20year%20on%20record%20%7C%20Copernicus

Copernicus Climate Change Service. (2024, January 9). *Copernicus: 2023 is the hottest year on record, with global temperatures close to the 1.5°C limit.* https://climate.copernicus.eu/copernicus-2023-hottest-year-record

Cunsolo, A., & Ellis, N. R. (2018). Ecological grief as a mental health response to climate change-related loss. *Nature Climate Change, 8*(4), 275–281.

Dance, S. (2023, November 9). You've just lived through the Earth's hottest 12 month on record. *The Washington Post Online.* www.washingtonpost.com/weather/2023/11/09/earth-hottest-months-climate-warming/

Davis, H., & Todd, Z. (2017). On the importance of date, or, decolonizing the Anthropocene. *ACME: An International Journal for Critical Geographies, 16*(4), 761–780. https://acme-journal.org/index.php/acme/article/view/1539

De Oliveira, V. M. (2021). *Hospicing modernity: Facing humanity's wrongs and the implications for social action.* Penguin Random House.

Derr, V., & Simons, J. (2020). A review of photovoice applications in environment, sustainability, and conservation contexts: Is the method maintaining its emancipatory intents? *Environmental Education Research, 26*(3), 359–380. https://doi.org/10.1080/13504622.2019.1693511

Doherty, T. J., & Clayton, S. (2011). The psychological impacts of global climate change. *American Psychologist, 66*(4), 265–276.

Donald, D. T. (2012). Forts, curriculum, and ethical relationality. In N. Ng-A-Fook & J. Rottmann (Eds.), *Reconsidering Canadian curriculum studies: Provoking historical, present, and future perspectives* (pp. 39–46). New York.

Donald, D. T. (2016). From what does ethical relationality flow? An "Indian" act in three artifacts. In J. Seidel & D. W. Jardine (Eds.), *The ecological heart of teaching: Radical tales of refuge and renewal for classrooms and communities* (pp. 10–16). Peter Lang.

Driscoll, S., Rubin, T., Campo, S., Grobeckker, K., Boisvert, M.-L., Wiseman, D., & Jao, L. (2023, March). *The impact of the Pandemic on student-directed STEM inquiry connections to the Quebec Education Program.* Poster presented at Bishop's University Research Week, Sherbrooke, QC.

Elliott, K. (2016). Caring masculinities: Theorizing an emerging concept. *Men and Masculinities, 19*(3), 240–259. https://doi.org/10.1177/1097184X15576

Ellis, J. I., Ellet, A. J., & DeWeaver, K. (2007). Human caring in the social work context: Continued development and validation of a complex measure. *Research on Social Work Practice, 17*(1), 66–76. https://doi.org/10.1177/1049731506293339

Favara-Scacco, C., Smirne, G., Schilirò, G., & Di Cataldo, A. (2001). Art therapy as support for children with leukemia during painful procedures. *Medical and Pediatric Oncology, 36*(4), 474–480.

Field, E., Schwartzberg, P., & Berger, P. (2019). *Canada, climate change and education: Opportunities for public and formal education.* http://lsf-lst.ca/media/National_Report/National_Climate_Change_Education_FINAL.pdf

Freilich, R., & Shechtman, Z. (2010). The contribution of art therapy to the social, emotional, and academic adjustment of children with learning disabilities. *Arts Psychother, 37*(2), 97–105. www.sciencedirect.com/science/article/pii/S0197455610000201

Gifford, R. D., & Chen, A. K. (2017). Why aren't we taking action? Psychological barriers to climate-positive food choices. *Climatic Change, 140*(2), 165–178.

Gill, J. (2023). The asteroid and the fern. In R. Solnit & T. Young Lutunatabua (Eds.), *Not too late: Changing the climate story from despair to possibility* (pp. 124–129). Haymarket Books.

Greenberg, M. T., & Harris, A. R. (2012). Nurturing mindfulness in children and youth: Current state of research. *Child Development Perspectives, 6*(2), 161–166. http://doi.org/10.1111/j.1750-8606.2011.00215.x

Greene, S., Burke, K. J., & McKenna, M. K. (2016). When words fail, art speaks: Learning to listen to youth stories in a community photovoice project. In *Youth voices, public spaces, and civic engagement* (pp. 247–270). Routledge.

Hayes, K., & Kaba, M. (2023). *Let this radicalize you: Organizing and the revolution of reciprocal care.* Haymarket Books.

Hayhoe, K. (2023, October 24). Hope in the climate change discussion [Video File]. *2023 Beck Environmental Lecture.* YouTube. www.youtube.com/watch?v=QL4bsOY2538

Henrich, J., Heine, S. J., & Norenzayan, A. (2010). The weirdest people in the world? *Behavioral and Brain Sciences, 33*(2–3), 61–83. https://doi.org/10.1017/S0140525X0999152X

Hmelo-Silver, C. E., Duncan, R. G., & Chinn, C. A. (2007). Scaffolding and achievement in problem based and inquiry learning: A response to Kirschner, Sweller, and Clark (2006). *Educational Psychologist, 42*(2), 99–107. https://doi.org/10.1080/00461520701263368

Jao, L., Wiseman, D., Di Placido, C., Ako, M., & Choi, S. J. (2023). *Five years in review: A research-practice partnership with trafalgar school for girls, Bishops University and McGill University* (20 pages). Trafalgar School for Girls.

Jao, L., Wiseman, D., & Loupelle, C. (2019, June). *The READ dress project: Student-directed STEAM inquiry.* Paper presented at the 2019 National Coalition of Girls' Schools (NCGS) Conference, Pasadena, CA.

Jardine, D. W., Clifford, P., & Friesen, S. (2006). *Curriculum in abundance.* Lawrence Erlbaum Associates.

Klocker, N., Gillon, C., Gibbs, L., Atchison, J., & Waitt, G., (2021). Hope and grief in the human geography classroom. *Journal of Geography in Higher Education, 47*(5), 737–754. https://doi.org/10.1080/03098265.2021.1977915

Lam, S., & Trott, C. D. (2022). Children's climate change meaning-making through photovoice: Empowering children to learn, care, and act through participatory process. *ESC, 62.* https://ojs.up.pt/index.php/esc-ciie/article/view/478

Lear, J. (2006). *Radical hope: Ethics in the face of cultural devastation.* Harvard University Press.

Léger-Goodes, T., Malboeuf-Hurtubise, C., Hurtubise, K., Simons, K., Boucher, A., Paradis, P.-O., Herba, C. M., Camden, C., & Généreux, M. (2023a). How children make sense of climate change: A descriptive qualitative study of eco-anxiety in parent-child dyads. *PLoS ONE, 18*(4). https://doi.org/10.1371/journal.pone.0284774

Léger-Goodes, T., Malboeuf-Hurtubise, C., Lefrançois, D., Gagnon, M., Herba, C. M., Paradis, P.-O., & Ethier, M.-A. (2024). La pertinence de la philosophie pour enfants dans la gestion de l'écoanxiété chez les enfants du primaire. *Éthique En Éducation et En Formation – Les Dossiers Du GREE, 15,* 99–124. https://doi.org/10.7202/1110004ar

Lévinas, E. (1998). *Entre nous: Thinking-of-the-other* (M. Smoth & B. Harshav, Trans.). Columbia University Press.

Liboiron, M. (2021). *Pollution is colonialism.* Duke University Press.

Long, J. S. (2011). Labelling angles: Care, indifference and mathematical symbols. *For the Learning of Mathematics, 31*(3), 2–7. www.jstor.org/stable/41319600

Malboeuf-Hurtubise, C., Di Tomaso, C., Lefrançois, D., Mageau, G. A., Taylor, G., Éthier, M. A., Gagnon, M., & Léger-Goodes, T. (2021a). Existential therapy for children: Impact of a philosophy for children intervention on positive and negative indicators of mental health in elementary school children. *International Journal of Environmental Research and Public Health, 18*(23). www.mdpi.com/1660-4601/18/23/12332

Malboeuf-Hurtubise, C., Léger-Goodes, T., Herba, C., Bélanger, N., Smith, J., & Marks, E. (2024). Meaning making and fostering radical hope: Applying positive psychology to eco-anxiety research in youth. *Frontiers in Child and Adolescent Psychiatry, 3.* https://doi.org/10.3389/frcha.2024.1296446

Malboeuf-Hurtubise, C., Léger-Goodes, T., Mageau, G. A., Joussemet, M., Herba, C., Chadi, N., Lefrançois, D., Camden, C., Bussières, È.-L. Geneviève, T., Éthier, M.-A., & Gagnon, M. (2021b). Philosophy for children and mindfulness during covid-19: Results from a randomized cluster trial and impact on mental health in elementary school students. *Progress in Neuro-Psychopharmacology & Biological Psychiatry, 107*, Article 110260. www.sciencedirect.com/science/article/pii/S0278584621000191

Marks, E., Atkins, E., Garrett, J. K., Abrams, J. F., Shackleton, D., Hennessy, L., Mayall, E. E. Bennett, J., & Leach, I. (2023). Stories of hope created together: A pilot, school-based workshop for sharing eco-emotions and creating and actively hopeful vision of the future. *Frontiers in Psychology, 13.* https://doi.org/10.3389/fpsyg.2022.1076322

McLarnon, M., Wiseman, D., Malboeuf-Hurtubise, C., Cognet, E., Equivel, L. R., & Léger-Goodes, T. (Forthcoming). Whose need is it anyway? Negotiating research roles, epistemologies and ethics in development of an interdisciplinary "needs assessment" for teaching and learning for climate change hope. In C. Mitchell, K. Schwarz & R. Hutton (Eds.), *Participatory data analysis in/as feminist research.*

Minner, D. D., Levy, A. J., & Century, J. (2010). Inquiry-based science instruction–what is it and does it matter? Results from a research synthesis years 1984–2002. *Journal of Research in Science Teaching, 47*(4), 474–496. https://doi.org/10.1002/tea.20347

National Oceanic and Atmospheric Administration. (2024, January 12). *2023 was the world's warmest years on record, by far.* www.noaa.gov/news/2023-was-worlds-warmest-year-on-record-by-far#:~:text=Earth%27s%20average%20land%20and%20 ocean,0.15%20of%20a%20degree%20C.

Noddings, N. (1984). *Caring: A feminine approach to ethics and moral education.* University of California Press.

Noddings, N. (1988). An ethic of caring and its implications for instructional arrangements. *American Journal of Education, 96*(2), 215–230.

Noddings, N. (2005). *The challenge to care in schools: An alternative approach to education* (2nd ed.). Teachers College Press.

Ojala, M. (2017). Hope and anticipation in education for a sustainable future. *Futures, 94*, 76–84. http://doi.org/10.1016/j.futures.2016.10.004

Ojala, M. (2012). How do children cope with global climate change? Coping strategies, engagement, and well-being. *Journal of Environmental Psychology, 32*(3), 225–233. https://doi.org/10.1016/j.jenvp.2012.02.004

Paddison, L. (2023, November 9). Humanity just lived through the hottest 12 month in at least 125,000 years. *CNN.* www.washingtonpost.com/weather/2023/11/09/earth-hottest-months-climate-warming/

Passmore, H. A., Lutz, P. K., & Howell, A. J. (2022). Eco-anxiety: A cascade of fundamental existential anxieties. *Journal of Constructivist Psychology, 36*(2), 138–153.

Pihkala, P. (2022). The process of eco-anxiety and ecological grief: A narrative review and a new proposal. *Sustainability, 14*(24), Article 24. https://doi.org/10.3390/su142416628

Robbins, P., & Moore, S. A. (2012). Ecological anxiety disorder: Diagnosing the politics of the Anthropocene. *Cultural Geographies, 20*(1), 3–19. http://doi.org/10.1177/147447401246988

Rouf, K., & Wainwright, T. (2020). Linking health justice, social justice, and climate justice. *The Lancet: Planetary Health, 4*(4), 131–132. www.thelancet.com/journals/lanplh/article/PIIS2542-5196(20)30083-8/fulltext

Rubin, T., Campo, S., Grobeckker, K., Boisvert, M.-L., Wiseman, D., & Jao, L. (2022, March). *Mapping student directed STEM inquiry on the Québec education programme: A school-university research-practice partnership.* Poster presented at Bishop's University Research Week, Sherbrooke, QC.

Rummery, K., & Fine, M. (2012). Care: A critical review of theory, policy and practice. *Social Policy & Administration, 46*(3), 321–343. https://doi.org/10.1111/j.1467-9515.2012.00845.x

Schuster, R., Germain, R. R., Bennett, J. R., Reo, N. J., & Arcese, P. (2019). Vertebrate biodiversity on Indigenous-managed lands in Australia, Brazil and Canada equals that in protected areas. *Environmental Science and Policy, 101*, 1–6. https://doi.org/10.1016/j.envsci.2019.07.002

Solnit, R., & Young Lutunatabua, T. (Eds.). (2023). *Not too late: Changing the climate story from despair to possibility.* Haymarket Books.

Stein, S., Andreotti, V., Suša, R., Ahenakew, C., & Čajková, T. (2022). From "education for sustainable development" to "education for the end of the world as we know it". *Educational Philosophy and Theory, 54*(3), 274–287. https://doi.org/10.1080/00131857.2020.1835646

Strack, R. W., Magill, C., & McDonagh, K. (2004). Engaging youth through photovoice. *Health Promotion Practice, 5*(1), 49–58.

Sultana, F. (2022a). Resplendent care-full climate revolutions. *Political Geography, 99*(November), 102785. https://doi.org/10.1016/j.polgeo.2022.102785

Sultana, F. (2022b). The unbearable heaviness of climate coloniality. *Political Geography, 99*(March), 102638. https://doi.org/10.1016/j.polgeo.2022.102638

Thompson, A. (2023, November 9). Earth just has the hottest 12-month span in recorded history. *Scientific American Online.* www.scientificamerican.com/article/earth-just-had-the-hottest-12-month-span-in-recorded-history/

Tronto, J. C. (1993). *Moral boundaries: A political argument for and ethic of care.* Routledge.

UNDP. (2023, February 9). How can we emerge stronger from today's multiple crises. *Oslo Governance Centre.* www.undp.org/policy-centre/oslo/events/how-can-we-emerge-stronger-todays-multiple-crise

UNESCO. (2019, December 16). *Strong education commitments at UN climate summit in Madrid.* https://en.unesco.org/news/strong-education-commitments-climate-summit-madrid

Ungerson, C. (2006). Gender, care, and the welfare state. In K. Davis, M. Evans & J. Lorber (Eds.), *Handbook of gender and women's studies* (pp. 272–286). Sage.

Voiland, A. (2023). Tracking Canada's extreme 2023 fire season. *NASA Earth Observatory.* https://earthobservatory.nasa.gov/images/151985/tracking-canadas-extreme-2023-fire-season

Waldby, C. (2012). Medicine: The ethics of care, the subject of experiment. *Body & Society, 18*(2–4), 179–192. https://doi.org/10.1177/1357034X12451778

Wang, C., & Burris, M. A. (1997). Photovoice: Concept, methodology, and use for participatory needs assessment. *Health Education & Behavior, 24*(3), 369–387

Wiseman, D., & Jao, L. (2019, June). *The READ dress project: Student-directed STEAM inquiry and change in practice.* Paper presented at the Canadian Society for the Study of Education Annual Conference, Vancouver, BC.

Wiseman, D., Jao, L., Rubin, T., Boisvert, M-L., Loupelle, C., Campo, S., & Grobbecker, K. (2022a, May). *Mapping student-directed STEM inquiry onto the Québec Education Program.* Paper presented at the Canadian Society for the Study of Education, online. https://openjournals.library.sydney.edu.au/STEMEC2022/article/view/15981/14514

Wiseman, D., Jao, L., Rubin, T., Grobbecker, K., Campo, S., & Boisvert, M.-L. (2022b, November). *Meeting curricular requirements: A mapping of student-directed STEM.* Paper presented at the 7th STEM in Education Conference, Sydney, Australia.

Wiseman, D., Lunney Borden, L., Beatty, R., Jao, L., & Carter, E. (2020). Whole-some artifacts: (STEM) teaching and learning emerging from and contributing to community. *Canadian Journal of Science, Mathematics, and Technology Education, 20*(2), 264–280. https://doi.org/10.1007/s42330-020-00079-6

Wiseman, D., Lunney Borden, L., & Sylliboy, S. (In press). STEAM as informed by netukulimk: Engaging in the radical to consider how to do things differently. In E.-J. A. Kim & K. Czuy (Eds.), *Relational education beyond the fort: Land-based STEAM* (pp. 51–70). Peter Lang.

Wynter, S. (2003). Unsettling the coloniality of being/power/truth/freedom: Towards the human, after man, its overrepresentation – an argument. *The New Centennial Review, 3*(3), 257–337. https://doi.org/10.1353/ncr.2004.0015

Part 2

Addressing Climate Anxiety in the Classroom

Progressing from the theoretical foundations laid out in Part 1, Part 2 of the book delves into creative and innovative approaches that educators can use to help students process climate anxiety within the classroom. It begins with an exploration of how solarpunk drama can serve as a powerful pedagogical tool. Petersen and Farrell emphasize the need to move beyond conventional climate change curricula, which often fail to engage with the social, political, and cultural dimensions of the crisis. Instead, they advocate for a solarpunk pedagogy that fosters affective, relational, and ethical engagement in both teaching and learning. Through drama-based approaches, students are encouraged to explore their emotions, uncertainties, and lived experiences, creating a space where they can imagine and develop new strategies for addressing climate change. The chapter culminates with the play Resurgence, which vividly portrays the complex mix of uncertainty, panic, and hope in a world affected by climate change.

Building upon the use of drama and creative pedagogical tools to engage students emotionally and ethically with climate change, in Chapter 9 Rueter Veiga examines how Theatre of the Oppressed (TO) can be utilized to confront climate anxiety. TO's participatory approach allows students to critically analyse oppressions, share knowledge, and harness their understanding to challenge systems that contribute to climate change. By engaging in TO, students can build trust, gain a sense of control, and address feelings of helplessness that often accompany discussions of climate change. Rueter Veiga illustrates the potential of TO with an example of a climate drama project for students, which includes interactive activities and reflective exercises to help them process their emotions.

Chapter 10 then turns to the Finnish educational context, where Grünthal and colleagues explore how environmental education, intertwined with children's and young adult literature, can enhance students' emotional literacy and environmental agency. They discuss how literature, especially within the Finnish subject "Mother Tongue and Literature," can be used to foster eco-emotional multiliteracy among students, helping them navigate the complex emotions associated with the climate crisis. The chapter also addresses the challenges posed

DOI: 10.4324/9781003494416-9

by declining reading skills and motivation, offering strategies to engage young readers with climate-related literature.

In Chapter 11, the focus shifts to a personal account from Erin Madon, a high school science teacher in Ontario, Canada. Madon reflects on her journey in integrating climate change education into her teaching practice. She recounts the challenges and successes she encountered while trying to balance the academic demands of the science curriculum with the need to address the psychological and emotional impacts of climate change on her students. The chapter highlights the strategies she developed to help students manage their climate anxiety and provides insights into the effectiveness of these approaches based on student feedback.

Finally, Caroline Hickman concludes Part 2 with a discussion on the emotional and mental health impacts of climate change on children and young people. Hickman argues that eco-anxiety is a normal and healthy response to the environmental crisis and advocates for educational approaches that help students make sense of their distress, find meaning in their experiences, and build collective support systems. The chapter emphasizes the importance of validating young people's feelings and using creative methods, such as outdoor education and youth theatre, to help them articulate and manage their emotions in a safe and supportive environment.

Throughout Part 2, the narrative underscores the importance of integrating imaginative, relational, and ethical approaches in climate change education. By doing so, educators can help students not only process their climate anxiety but also transform it into a source of agency, resilience, and hope.

8 Solarpunk Pedagogy in a Warming World

Cultivating Radical Hope for an Uncertain Future

Christie Petersen and Alysha J. Farrell

Positionality Statement

Situated in a Canadian prairie province within Treaty 1 Territory, in the Global North, I (Christie) come to this work as a teacher in a metro K to 12 public school system, and as an early career researcher interested in the experiences of early career teachers and their Becoming in an uncertain world. Becoming teacher is an ongoing entanglement with myriad discursive, affective, and material forces, such as the climate crisis and climate anxiety, that complicate the work of learning to teach. The climate crisis is an urgent reality that pounds at the door of every school across the globe but strikes with greater immediacy in some countries more than others through catastrophic events such as uncontrollable fires, drought, famine, and extreme temperatures. Engaging with relational psychoanalysis and ecofeminist thought, I seek to explore how early career teachers are making sense of the global realities that are encountered in the work of teaching.

Located in Treaty 2 Territory, the shared lands of the Cree, Anishinaabeg, Anishininew, Dene, and Dakota Peoples, and the homeland of the Red River Métis Nation, I (Alysha) come to this work as a teacher educator and a researcher who studies the emotional impacts of the climate crisis in the lives of young people. I understand climate anxiety as a reasonable response to a psychological stress brought about by a real threat and not as a malady in need of a cure. The young people I have worked with over the years have taught me that they need significant others within relational processes to channel their environmental concerns into positive action. If their fears are ignored, it can result in debilitating anxiety and depression. That said, educators have an ethical responsibility to help young people process their climate anxiety from a place of radical love and determination. Drama-based pedagogies can be effective methods to work with young people's specific climate fears, foster healing connections, cultivate solidarity, and promote eco-resilience in schools.

> "We're *solarpunks* because the only other options are denial or despair."
> Adam Flynn, 2014, *Solarpunk: Notes toward a manifesto*

DOI: 10.4324/9781003494416-10"

It Is Not the Monsters Under the Bed Keeping Them Up at Night

Young people are scared. Their ability to envision a livable future flickers and fades further from view as they bear witness to the unethical in/actions of authority figures who worship at the altar of petro-capitalism. The Intergovernmental Panel on Climate Change (2022) warns that climate impacts on natural and human systems are becoming increasingly intense, frequent, and irreversible. These consequences extend beyond the environment; they affect mental health and elicit intense negative emotions (Farrell, 2022a). A recent survey of 1000 Canadian youth, aged 16–25, confirms that many young people are feeling afraid (66%), anxious (63%), helpless (58%), and powerless (56%) in the face of the climate crisis (Galway & Field, 2023, p. 3). These numbers are certainly cause for concern, yet experts in the burgeoning field of climate psychology warn educators against pathologizing the strong climate emotions of their students. Fear, intergenerational anger, grief, and anxiety are healthy responses to the danger young people are facing (Climate Psychology Alliance, 2022; Davenport, 2017; Weintrobe, 2013, 2021). What is unhealthy, however, is the pervasive silencing of the climate crisis in schools and the negation of the presence of climate anxiety that exists within students (Farrell, 2023).

The Learning for a Sustainable Future's *Survey on Canadian Perspectives on Climate Change and Education* (2022) reports that 67% of Canadians think schools need to give climate change education a higher priority, yet only 13% of the educators surveyed taught 11 or more hours of climate change content over the school year. We believe the continued suppression of the climate emergency, and subsequent climate anxiety, inflicts a moral injury upon students. It is also an abdication of the education system's responsibility to safeguard the mental and physical well-being of students. Our chapter focuses on enacting solarpunk pedagogy to open space in classrooms for the emotional composting of the climate crisis through the acknowledgment and recognition of climate anxiety. We begin with a description of solarpunk as a literary and aesthetic movement and then describe how solarpunk pedagogues are suited to the work of emotional composting with students. To illustrate the power of embodied learning in processing climate anxiety, we include a 10-minute solarpunk play script called *Resurgence*. The chapter concludes with ways that solarpunk praxis can help teachers and students process climate anxiety, foster radical hope, cultivate eco-identity, and bring more sustainable ways of living to life.

Setting the Stage

In 2021 and 2022, Alysha worked with 36 young people (ages 13 to 21) in southwestern Manitoba to better understand the emotional impacts of the climate crisis in their lives. The findings from her study indicate that when eco-anxiety is left unprocessed, it results in feelings of dread, intergenerational anger, and a diminished future orientation:

It feels like . . . fighting a battle you already lost before you are born.
I want to have hope, but I don't. I feel very defeated. And that is paralyzing.
What you strive to be when you were younger can't be that possible right now.

Figure 8.1 shows a small group of participants in Alysha's study who are engaged in what she would come to understand as emotional composting. In this workshop, the participants created a theatrical tableau to represent their feelings about the roots of the climate crisis. The creators of the tableau, along with some of the other participants, expressed strong feelings of climate anxiety:

- I see all these problems and . . . I don't know personally what I can do. It's just so stressful.
- People don't like to acknowledge things that scare them. [They] are scared of what they may be doing but don't wanna recognize it and don't really want to take action. It could feel like you're screaming at a wall and like no one's listening to you.

Figure 8.1 The Roots of the Climate Crisis

In response to the expressions of climate anxiety, Alysha invited the other participants in the workshop to adjust the tableau (proximity, facial expressions, posture, or shape) to build a new image of what it would feel like to live in a climate-aware school community. The original tableau, and the process of shifting it, opened a space for difficult conversations about the relationship between the roots of the climate crisis and a lack of reverence for people and other animals. More significantly, the participants reported feeling less anxious because of their participation in the drama exercises.

Contemporary Landscapes of Education

Within the unpredictable and volatile contemporary world, classrooms are becoming increasingly more complex as teachers are called upon to grapple with living (and teaching) in an uncertain world that is embroiled in a state of climate emergency. Teaching and learning in schools continues within disjointed relationships to the climate crisis (Farrell, 2022a). In classrooms, the reflections of a world that perhaps never was, and the radiance of an imagined world that might not be, blind teaching and learning to the mounting realities of the present. The looming shadows of climate crisis evoke affective responses and embodied reactions that become obscured in the shadows of "big data" and its attempt to cognitively rationalize, bracket, and compartmentalize the global emergency as a problem separate from humans (Haraway, 2016; Verlie, 2021). Teachers have an ethical responsibility (one could argue, an obligation) to students to take up the reality of the world as it is rather than the phantasy of a world that might be, while acknowledging the history of a fleeting world once known. Haraway (2016) characterizes this idea as *staying with the trouble*. Staying with the trouble, "requires learning to be truly present, not as a vanishing pivot between awful or edenic pasts and apocalyptic or salvific futures, but as mortal critters entwined in myriad unfinished configurations of places, times, matters, meanings" (p. 1). Climate anxiety is understood to be "difficult knowledge" (Pitt & Britzman, 2003) as it can intermingle, interfere, impede, and inspire. Teaching with climate anxiety imposes demands for relational spaces that have the potential to be transformative, generative, and rooted in possibilities – sustained by "radical hope" (Orange, 2017). Verlie (2021) warns that, "climate change is excessive, boundless and dizzying, and human experiences of this can be unfathomable, indescribable, and mysterious" (p. 56). The climate crisis is an emergency that can no longer be set aside to the periphery for a future generation to tackle, perpetually existing as "someone else's problem" (Klein, 2014). Teaching within this space and time means engaging with the world we are in and teaching in ways that reflect that world by staying with the trouble at hand (Haraway, 2016). Recognition and acknowledgment of encounters with climate anxiety can allow for the affective, material, and embodied lived experiences to be conceptualized within ethical and relational spaces that create possibilities for radical (re)imagination (Orange, 2017).

Disrupting Curriculum

Curriculum resists rigid definition and is largely understood to be far more expansive than a physical document outlining compartmentalized subjects through prescriptive outcomes. Here, curriculum is understood to be a, "complicated conversation between teachers and students over the past and the future and their meaning for the present" (Pinar, 2019, p. 1). Despite the complex and layered conceptualizations of curriculum (often situated in the field of curriculum studies), curriculum frequently reflects, "neoliberal and capitalist narratives of mastery and preparedness," and perpetuates notions of control, certainty, and guaranteed futures within teaching and learning (Tomin, 2020, pp. 2–3). Rigid and reductive approaches to curriculum negate the complexities and contradictions that are inherent to existence on this planet, turning away from the lived intensities of climate crisis. Recognizing the climate crisis as an entanglement of social, political, historical, and cultural phenomena opens up possibilities for recognizing the cognitive, physical, and emotional dimensions that manifest in climate anxiety (Bryan, 2020).

Climate Crisis as Entanglement

Comprehending climate crisis as an ethical and relational entanglement, connected to the past, present, and future, is to acknowledge, recognize, and bear witness to the intersecting systems of oppression that affect the human and more-than-human world (Adams & Gruen, 2022, p. xxi). Distinct from an environmental or sustainability lens, a critical theory lens, such as ecofeminism, considers the interconnectedness and entanglement of the human and the more-than-human world, while deconstructing and disengaging from hierarchies or dualisms that uphold patriarchal, capitalist, and neoliberal systems of power. There is an unsettling of anthropocentrism and decentering of humans so as not to privilege their interests over those from the more-than-human world (Adams & Gruen, 2022). Understanding climate crisis as a human produced phenomenon allows for the recognition that, "one of the major challenges of the climate crisis is that the enemy is not a virus, an organism external to us. We are our own enemy" (Ruitenberg & Rathje, 2022, p. 74). The climate crisis is a global emergency experienced through embodied and affective responses that can overwhelm cognitive reason. Nuanced, complex, and contextual, the climate crisis is entangled with social injustices, systemic racism, colonization, and slavery (Orange, 2017). Not merely a quantifiable scientific phenomenon, it is experienced in the body and in the mind as a gripping collective trauma that may cause a turning away from the present emergency and the overwhelming reality of an uncertain future (Orange,

2017). How might educators make space for students to contend with the present realities while imagining possibilities for futures that resist escape or surrender?

Speculative Fiction

Recognition and acknowledgement of a world in peril, a future uncertain, demands responses within the spaces of education that attend to, and contend with, the complexities that are produced by environmental crises. The tensions, uncertainties, and dissonance elicited by the material, embodied, and affective realities of climate crisis require curricular and pedagogical spaces that stretch beyond the empirical or objective desires often associated with quantitative research within the sciences. Rather, "arts-based research methods connect the material world with research participants' embodied experiences . . . [and] aim to bring performativity, psychodynamics, materiality and embodiment together" (Tian, 2023, pp. 7–8). The climate crisis, as an entanglement, necessitates educational spaces, and pedagogical moves, that welcome critical engagement, contemplation, and creativity. Spaces that will exist to invite embodied and affective modes for making sense of the world in ways that resist finite endpoints or tidy resolutions. Pedagogical decisions that generate hospitable spaces are inclusive, flexible, and supportive, open to exploration and provocation for ways of knowing and being in the world that invites the unimagined to emerge and take shape.

Speculative fiction and science fiction are storytelling genres (Tomin, 2020) that can make space for teachers and students, as a community, to struggle with the realities of contemporary issues, such as climate crisis, while creating hope (Agee, 2023) to imagine futures that offer greater possibility for the human, and more-than-human, world to exist in a way that does not compromise the possibility of life on this planet. Science fiction offers a unique entry point into thinking about the present moment and envisioning myriad possible futures. Whether it is imagining what other galaxies might hold through stories of intergalactic travel, exploring how changing the way we think about gender, race, sexuality, and other aspects of identity could impact future societies, or considering the effects that ecological decisions could have on humanity, science fiction can offer students an opportunity to think differently about contemporary issues and the part they plan in enacting change. (Tomin, 2020, p. 4)

The imagined worlds of speculative fiction create the space and safety to take risks in exploring radical alternatives that embrace difference and inclusion because there is no right or wrong answer; rather, there is the openness to re/configure and re/define what society might become and how we might live together in such a world (Tomin, 2020). Systems of education are complicit in their role of re/producing societal structures and conditions that appear to exist without alternatives, thereby foreclosing possibilities for the future. Pedagogical choices that seek out ethical, inclusive, and relational alternatives to the dominant discourses of contemporary society allows for negotiating, disrupting, conceptualizing, and imagining other possibilities of existence between the human and more-than-human worlds. The intention, "To defamiliarize the familiar is a key role of speculative fiction, and one thanks to which we can rediscover the world and its inhabitants" (Dědinová et al., 2021, p. 13). Speculative fiction, science fiction, and solarpunk provide spaces for possibilities that are not yet imagined and futures yet to be considered.

Solarpunk

Within science fiction, there are an array of subgenres that are distinguished from one another by certain values, perspectives, energy sources, and intentions. Solarpunk is a literary sub-genre of science fiction that is premised upon community, inclusivity, and sustainable futurity. Rather than imagining a world that is free from responsibility or concern for the planet such as through escapism, solarpunk imagines a future that has been impacted by the consequences of climate crisis and seeks out ways to live in community with the human, and more-than-human, world with the aim of making life, "better for all" (Wagner & Wieland, 2022, p. 1). Solarpunk embraces, "hope, resistance, and transformation" (p. 1), by imagining worlds that contend with and address the contemporary environmental crises while striving to cultivate a society that considers all on this planet. The embrace of counternarratives creates space to imagine possibilities for a future that allows life on Earth to exist through the disruption of the taken-for-granted ways and oppressive social constructs that dictate contemporary existence (Agee, 2023).

Solarpunk's Historical and Theoretical Influences

Solarpunk, a type of speculative fiction within the science fiction genre, initially garnered attention between 2008 and 2014 through online presence and publications. Originally, solarpunk was characterized by community, green spaces, and art nouveau designs (Wagner & Wieland, 2022). Despite being influenced by cyberpunk and steampunk, solarpunk remains distinct in its aspirations for worlds that are sustainable rather than worlds of utopian or dystopian landscapes (Wagner & Wieland, 2022). Solarpunk contemplates the environmental concerns of the present through imaginings of futures that enact change to avoid extinction and embrace other ways for cultivating existence within community. The classification of solarpunk has been used retroactively to characterize literary works that were published prior to the inception of the genre in the early 2000s. Regarded as a social and political movement, solarpunk seeks to imagine alternatives to capitalism, the commodification of natural resources, and systems of oppression. The genre of solarpunk is open to, and welcoming of, myriad ways of being and knowing through its inclusive approach to community and alterity. Solarpunk is not narrowly defined or prescriptive; rather, this genre generates spaces for imagining possibilities of futures that are hospitable to all, rather than some. Attention is critically directed to the interconnectedness of the world as solarpunk seeks to deconstruct, in theory and in action, oppressive and devastating systems of capitalism, white supremacy, consumerism, and colonialism in order to imagine an inclusive community of care for all (Wagner & Wieland, 2022).

Pedagogy

Pedagogy encompasses the methods and practices of teaching that are informed by, and reflective of, assumptions, values, and beliefs (Breault, 2010). Pedagogy is defined in contrast to instruction as instruction is often considered to be context-free, whereas pedagogy is informed by the contexts of the lived world (Breault, 2010). Solarpunk pedagogy is infused with a concern for all to live as part of a community that is welcoming of difference. In thinking with solarpunk, attention is drawn to the entanglement of climate crises that creates opportunities to explore with students the oppressive systems (e.g. colonialism, patriarchy, white supremacy) that contribute to and perpetuate climate change (Bryan, 2020). Solarpunk pedagogy,

> Purposefully engages educational communities in dynamic, multiple, pluralistic conceptions of the future that can help students and teachers alike process potential disaster, imagine alternatives, envision potential actualization of preferred futures, and live meaningfully with the uncertainty and unknowability of a future that exists transiently and always just out of view.
>
> (Tomin, 2020, p. 4)

A mirror is brought to the contemporary lived realities of climate crisis and allows for imaginative play with reflections so as to alter reality as we imagine other futures.

Affective, Relational, and Ethical Pedagogical Encounters

Curriculum and pedagogy are often treated as insular and isolated entities that retain a neutrality, remaining impervious to social, political, or cultural influence. Curriculum studies help us to recognize curriculum and pedagogy as contested spaces within education, saturated with the values, beliefs, and agendas of those with power and voice. Curriculum and pedagogy are the result of choices made regarding which knowledge is included, which knowledge is prioritized, and who gets to make those decisions (Apple, 2018). When curriculum and pedagogy are not recognized as more than physical documents or technical practices, the impacts and consequences to teaching and learning may go unacknowledged. Within these discussions, it is imperative to critically consider the larger purposes of education and whether or not we are permitting contemporary lived realities to be brought into the conversations within educational spaces. Teaching is a relational, ethical, and affective encounter with others and within a world that is ongoing. Pedagogy that recognizes and acknowledges the affective, relational, and ethical threads within lived experience provides expansive and generative spaces for making sense of a world that is complex, contradictory, and often confounding. Attunement to the affective, relational, and ethical dimensions of the lifeworld allows for recognition of the interconnectedness and draws into nearness that which can often feel invisible but simultaneously inescapable and incomprehensible.

Solarpunk Pedagogy

Infusing pedagogy with the guiding notions of solarpunk allows for ethical, relational, and affective engagements with teaching and learning. Solarpunk turns towards the lived realities of the world to critically consider how imagined futures may allow for possibilities to address the climate emergencies in ways that care for all members of this planet's community – human as well as the more-than-human. Pedagogy that embraces the aspirations of solarpunk creates opportunities for disruption of the normalized and taken-for-granted ways of knowing and being. The power of imagination and storytelling, through the perspective of solarpunk, wrestles with the tenuous places in knowledge and identity to nudge and provoke the cracks within the social fabric and produce openings that explore other ways of knowing and being that resist pre-determined or rigid definition.

Emotional Composting With Solarpunk

Solarpunk is an aesthetic and literary movement that imagines how we might enact a sustainable future that works for all humans and the more-than-human world (Wagner & Wieland, 2022; Więckowska, 2022) and posits radically different economic and social structures in which power is decentralized so as to live communally. Without shying away from the grittiness of making substantial changes to how we live, solarpunk offers a regenerative vision for the future. Justice and compassion are integral values in this genre, making solarpunk praxis a source of radical hope and a creative space to develop concrete strategies to build a better world. While resisting sentimentality, this sub-genre offers an antidote for the dread many young people feel when seemingly little can be done on an individual level to fix the climate crisis. As Agee (2023) aptly notes,

> Dystopian stories surely have a place, as a warning, but sometimes I feel like I've been warned enough. I want to know what to do in the face of despair, to not only avoid being crushed, but to reach for brighter skies.
>
> (p. 26)

As educators and researchers in the field of education, we are embedded in the same climate crisis as our students. Operating as a generative and communal guerilla-type therapy in classrooms, we assert that solarpunk pedagogies can help teachers and students collectively bear the unbearable knowledge of the climate crisis. Solarpunk allows for a deconstruction of contemporary realities, drawing insidious and oppressive systems out of the shadows, to imagine – and possibly action – radical alternatives for all human and more-than-human inhabitants of this earth. Additionally, solarpunk pedagogies welcome teachers and students to not only portray but contend with the tenuous and contradictory responses to climate crisis (Tanner, 2021). The embodied and psychological effects of climate crisis strike, resonate, and proliferate in the emotional manifestations of climate anxiety and grief (Bryan, 2020). The emotional dimensions of engaging with climate crisis in the classroom require care and attention for the unanticipated and difficult reactions that may be evoked in students' recognitions of the magnitude of the present-day threats to their existence, their complicity, and the inevitable tarnish of a once expected future. Rather than wallow and acquiesce to climate crisis as though it is a fait accompli, solarpunk generates spaces to deconstruct, mess around, and re/imagine radical alternatives that acknowledge the repercussions of climate change but embrace a hope for living with the world in a different way. This recursive, messy, and non-linear "process" can be understood as *emotional composting*. Emotional composting is the conversion of powerful feelings into somatic learning. It requires a shared experience that involves trusted others within an interpretive space to power this co-regulatory process. It calls people to use their bodies or other media, to express multifaceted feelings to others about emotionally charged information or events. Emotional composting heightens intuition, builds empathy, and fosters resilience in the face of uncertainty and stress. It allows us to re/conceptualize how the emotional dimensions of climate crisis are un/acknowledged, named, and explored in ways that are productive and generative in order to begin to address the complicated and difficult emotions that are generated by existing in and with climate crisis. Through emotional composting, teachers and students may come to recognize that seemingly negative emotions have the capacity to draw our attention to the injustices of the world and give rise to radical alternatives that break out of the confines of binary thought that oversimplifies and negates the tenuous complexities of living with climate crisis (Ahmed, 2010; Tanner, 2021). Solarpunk pedagogies create spaces for emotional composting so that students have opportunities to confront the overwhelming and confounding emotions of living in a world where existence is not definite, and uncertainty abounds. Drama-based pedagogies construct conceptual spaces and places for students to deposit resonances of their climate anxiety, emotions, and grief, draw their emotions into imaginary worlds, and, through emotional composting, begin to cultivate the seeds for radical alternatives that have the capacity to grow into action.

Writing the short play became an emotional composting process for us through solarpunk praxis. It amplified our own biophilic impulses, created an opportunity for us to reflect on our own eco-anxiety, and generated some new ideas on how to engage students with the difficult knowledge of biodiversity loss and the climate crisis, with care, courage, and humility. Playwrighting created a transitional space to cultivate a custodial ethic, one that assumes, "we are not only in the world but with the world, not out of the world or above it but with it" (Thornton et al., 2019, p. 240). Articulated in the Association of Canadian Deans of Education (ACDE) Accord on Education for a Sustainable Future (2022), education has a critical role to play in cultivating our eco-identity, shaping our inner dialogue which informs if and how we act in the face of the climate crisis, raising awareness about the inequitable impacts of climate change, and caring for others, particularly when resources are scarce (Farrell, 2022b). As solarpunk pedagogues and researchers, we call for the rewilding of curriculum and pedagogy, space for the emotional dimensions of the climate crisis in classrooms, and compassionate co-existence as a central aim of education in the Anthropocene.

Drama-Based Approaches to Pedagogy

Arts-based pedagogies open a liminal space for students to freely express themselves (Pihkala, 2018), empower and transform their thoughts and feelings, and encourage pro-environmental actions (Rousell & Cutter-MacKenzie-Knowles, 2020) through open-ended, creative, embodied, experiential, and participatory methods (Bentz, 2020; Trott et al., 2020).

Drama-based approaches to pedagogy provide imaginative spaces that create productive and generative possibilities for engaging with lived experiences, emotions, and uncertainties. The interpretive nature of drama lets contemporary realities be drawn into imaginative spaces to shape and cultivate alternative understandings. Drama-based pedagogies, such as script writing and plays, allow for a deconstruction and re/consideration of the extra/ordinary so that the taken-for-granted aspects of lived experience are drawn into the foreground. The imaginary and creative spaces of drama welcome competing perspectives to reside within the same space or scene without seeking resolution. The tensions are permitted to push and pull in ways that provoke, disrupt, and surprise through spontaneous and organic movements. Drama-based pedagogies create safe spaces and places to express emotions through characters and dramatic encounters that encourage risk-taking, while allowing opportunities to cultivate empathy as alternative perspectives are embraced to see complex issues through new eyes. Many drama-based experiences are collective and collaborative, embodying relational approaches that necessitate nuanced attention and response to one another. The imagination required of fictional world building calls upon creative problem solving that is not satisfied by one-dimensional decisions or choices. As a collective experience, drama makes space to imagine and welcome other ways of living, knowing, and being that are tethered – loosely or tightly – to the present world. The affective dimensions and strong emotions of lived experience, often relegated to private spaces, are embraced within drama-based pedagogies that recognize the simultaneously cognitive and emotional facets of existence in this world. We hear firsthand from a teacher who participated in an arts-based climate crisis research project on the impact of observing their students engaging in drama-based pedagogical experiences:

> Having witnessed my students' experiences and (having taken part myself) I am less likely to retreat into denial and escapism or be reluctant to take students into this terrifying territory for fear of hurting them. They are going there anyway and, at least, we can go together. I have seen that when we express our thoughts and feelings to and with others who share our pain, we can actually survive the feelings and find ourselves empowered by not being alone, and by being allowed to speak of the climate crisis without being shut down by those who would rather not hear it.

Within this teacher's account of utilizing drama-based pedagogies to engage with the incomprehensible but deeply felt experiences of climate crisis, we hear how hope and action can be cultivated from climate emotions of grief, despair, denial, shame, and guilt.

Resurgence: A Solarpunk Script

The following play is an example of a speculative fictional space in the genre of solarpunk. Titled *Resurgence*, this play offers moments of uncertainty, panic, and hope for finding ways to live in a warming world. It emerged from our ruminations in the participants' expressions of climate anxiety and the visceral realization that it is wholly insufficient for teachers to open space for expressions of climate anxiety without engaging students in the work of *emotional composting*.

The opportunity to narrate the overwhelming uncertainty that is part of climate crisis opens up to consideration of the complexities and competing notions that complicate human and more-than-human life. Within the possibilities that are cultivated in drama-based pedagogies, we find ways to contend with and challenge the taken-for-granted constructs of society so that we might seek out radical alternatives through the momentum of action fuelled by hope. "Theatre and performance can offer new frames of thinking, feeling, and viewing, or tell/show us something about our current ecological situation" (Woynarski, 2020, p. 2). We hope the characters in *Resurgence* inspire teachers and students to engage with the shadow of the climate crisis in ways that are generative.

Resurgence

Setting:	Set against the backdrop of a tiny solar-powered home. There is a large plant (ARIA) centre stage. The lighting is ethereal. Two old chairs are placed upstage left. Hanging over one of the chairs is an old blanket.
Characters:	BEATTIE: 87 years old, teacher educator, wearing a kaftan ZEPHYR: Teacher educator ARIA: Bio-luminescent plant, healer
At Rise:	Beattie sings softly to Aria. As Beattie sings, Aria responds.
ZEPHYR:	(Enters running from upstage left.) Am I too late?
BEATTIE:	(Holds up her hand to signal Zephyr to be quiet. Beattie's body sways as she delights in her interaction with Aria.)
BEATTIE:	(To Aria.) All in good time, old friend.
ZEPHYR:	(Walks slowly towards Beattie. Leans in to scrutinize Aria.) Do you think it could be today?
BEATTIE:	Patience, Zephyr.
ZEPHYR:	(Reaches out to touch Aria.)
ARIA:	(Reacts to Zephyr's touch.)
ZEPHYR:	How can you stand it, Beattie? You've been waiting for 30 years! (Reaches out again to touch Aria.)
ARIA:	(Reacts strongly to Zephyr's movement).
ZEPHYR:	(Jumps back astonished.) Thigmotropism. A chemical response to environmental stimuli . . . Fascinating.
BEATTIE:	(Gestures to Zephyr and Aria.) You can't distill this marvellous dance to a mechanosensory response!
ZEPHYR:	Oh, this is a dance, is it?
BEATTIE:	(While shaking her hips). Bada boom. Bada boom. Bada boom, boom, boom! (Stumbles a bit. Grabs onto ZEPHYR's arm to steady herself.)
ZEPHYR:	Here. (Guides Beattie to hold onto Aria.) Hold on tight for a minute. (Goes and gets one of the chairs for Beattie.)
ZEPHYR:	(Helps Beattie into the chair. Pours her a cup of tea.) Now tell me, so I can regale future generations of students with stories about the greatest friendship of all time, how did Aria come into your life?
BEATTIE:	On the day I graduated, my professor presented me with a blue satin pouch. Inside was a precious seed. She told me there would

come a time when I would plant the seed or gift the choice to another teacher candidate who embodied the finest qualities of what it means to be an educator. Five generations of colleagues gifted the seed to their most promising students.

ZEPHYR: Well, beloved adviser to hundreds of teaching apprentices, you never encountered a teacher candidate worthy of such a gift? And before you answer, know that I'm preparing to take great offence as one of your former students.

BEATTIE: (Playfully taps Zephyr's arm pretending to scold Zephyr. She takes a sip of tea and looks straight ahead becoming a little more solemn.)

The year before the Green Revolution, the drought wiped out the winter wheat and the Water Wars intensified. COVID-33 ravaged communities and compassion was in short supply. The vitality of classroom life was subsumed by nightmares on fire and thirsty little mouths.

ZEPHYR: Broken systems manufacture scary dreams.

BEATTIE: They do. . . . During one particularly smokey day, when the ventilation system was having difficulty keeping up, I abandoned my class mid-lecture and headed home. In a frenzy, I scavenged an old pot and some viable soil. For an entire year, I shared my water rations with her, hoping every day she would sprout.

ZEPHYR: (Moves to Aria and scrutinizes her closely.)
You've lived together for a long time.

BEATTIE: Now tell me what you're on about with those wonderful teacher candidates of yours.

ZEPHYR: These days, The Ministry is focused on stripping what they refer to as "soft skills" from curriculum.

BEATTIE: How are you helping your students outrun such reductive thinking?

ZEPHYR: I invited them to think about teaching as an act of guerilla gardening. Subversive, I know.

BEATTIE: Tell me more.

ZEPHYR: The focus of this kind of teaching is to care for the unloved others. It's about empowering students to tend to the neglected or abandoned practices that once contributed to the flourishing of life. So, we read voraciously to seed-bomb our pedagogies and when necessary, we engage in the clandestine cultivation of curriculum to promote environmental stewardship and connectivity to place.

BEATTIE: This is beautiful, emotional, and risky work.

ZEPHYR: Challenging for sure. Fleets of consultants are framing this type of orientation to teaching as a malady in need of a cure.

BEATTIE: Beware the deadpan technocrat. Always trying to make the work sterile. Terrified of unwieldy emotions. As if it's possible for the

	young to dislocate from ancestral howls of prolonged droughts, food shortages, wildfires, and super storms.
ZEPHYR:	To come at it slant, I've been talking about a healing orientation in curriculum making as opposed to a fixing approach.
BEATTIE:	What brought you and your students to the topic?
ZEPHYR:	The recent news frenzy about the Institute's interest in Aria.
BEATTIE:	Ahhh, those researchers are in cahoots with their corporate sponsors. (Gets out of chair and moves over to Aria and Zephyr.)
ZEPHYR:	Her nectar purifies water. The leaves, brewed into a tea, act as a detoxifier. And the bloom, every 30 years or so, is rumoured to –
BEATTIE:	The Institute wants to control her. Own her.
ZEPHYR:	They want to learn from her. It's not like in the old days. There are rules now. In fact, they've got a whole new wing dedicated to cross-species care and communication. If you give her to the Institute, Aria will be a part of a world-class medical research programme -
BEATTIE:	She's not mine to give. Aria's life is an end unto itself. If she's forcibly removed from her home and taken to the Institute, we all move further down the path of pernicious anthropomorphism. We do not have to extract to exist.
ZEPHYR:	This isn't your grandmother's extractivism. It's harnessing the medicinal properties of a renewable resource for the betterment of society.
BEATTIE:	Who is included in your society? We are in Aria's world, not outside of it.
ZEPHYR:	Now I've gone and upset you. What can I do?
BEATTIE:	Protect Aria's right to stay rooted in this place. Recognize that our lives are nested together. Help me keep my promise.
ZEPHYR:	Beattie, please. You're not going to be around forever.
BEATTIE:	In fact, I'm not going to be around tomorrow.
ZEPHYR:	What do you mean? Where are you going?
BEATTIE:	If I knew the answer to that I'd be famous.
ZEPHYR:	You're not making sense.
BEATTIE:	Then let me be clear. Today is the last day I walk the Earth.
ZEPHYR:	Wait a minute. Wait just a minute.
BEATTIE:	Now settle down. (Leads Zephyr to sit in her chair.) You already know that I've got one foot in the other side of this world. Bloody hell. I'm 87 years old!
ZEPHYR:	How do you know? I mean, how do you know you're going to . . .
BEATTIE:	Die today? You need to practice saying the D word.

ZEPHYR: How do you know you're going to die today?
BEATTIE: Intuition. Irregular heartbeats. Laboured breathing. It's in the tea leaves. An alchemy of signs.
ZEPHYR: Aren't you afraid?
BEATTIE: A little.
ZEPHYR: (Gets up in a hurry. Takes a small tool from a pocket.)
             What if Aria could help! Just a few sepals.
BEATTIE: (Walks over to Zephyr and Aria.)
             Help with what and at what cost? If you remove any number of sepals, you could interfere with the immanent bloom. And that could be disastrous for our friend.
ZEPHYR: The sepal tea might give you a little more time. After all the years you protected her, she can't do this small thing for you?
BEATTIE: You cannot, even with the best of intentions, exclude Aria from the realm of ethics. If you take from her, something she's not ready to give, it will cause harm.
ZEPHYR: Explain to me why harvesting a few leaves is such a moral calamity.
BEATTIE: In denying Aria her share of the Earth, you sever a tie in the more-than-human world. And reducing her to a medical tool prevents you from engaging her as an active participant in her own life.
ZEPHYR: You must want to go on for a little bit longer?
BEATTIE: I've had an extraordinary life. I've seen young teachers stand firm in the face of increasing pressure to colonize their creativity. I've been blessed to witness the hearts of students' catch fire when they experience the curative effects of cross-species care. My God, the generosity and compassionate wisdom I've received from those who understand teaching as an act of radical love . . . When the sun goes down tonight my good friend, I will die sated with life.
ZEPHYR: (Helps Beattie back to her chair.)
BEATTIE: You and me, two wings of the same bird. Now come and sit. We can't add more days to my life, but we can add a smidge more life to today.
ZEPHYR: (Gets the other chair and sits so that Aria is between them.)
BEATTIE: I'm really going to miss watching the sunflowers chase the sun.
ZEPHYR: What if I'm not ready to let you go?
BEATTIE: Beautiful things grow wild at the edges of our influence. As teachers, we must learn this lesson over and over.
             (Several beats of silence.)
ZEPHYR: Beattie?
             (Gets up and places the blanket over Beattie's lap.)
             Thank you for being my teacher.

Resurgence offers the audience a glimpse into a fictional future world that is neither a resplendent utopia nor a luminous paradise. Despite the context of an imagined world, the play is tethered to the contemporary realities of climate crisis through an extrapolation of thought that considers what might be to come. The fictional world of *Resurgence*, inclusive of human and more-than-human beings, contends with the perilous realities of water shortages, compromised food supplies, environmental crises, and viral pandemics (all of which presently complicate our current existences). We encounter Beattie's conversation with Zephyr about the subversive means by which teacher educators are seeking out ways which to educate teacher candidates through ethical and relational approaches that promote sustainability. We are welcomed into the friendship between Aria and Beattie, a relationship between a member of the human world and a member of the more-than-human world, to consider what it may mean to build a relationship between species. Through Beattie's conversation with Zephyr, we learn of the ways in which Beattie regards Aria, a bio-luminescent plant, as a friend and as a member of society that should remain free from capture for study or commodification. Beattie's recollection of a particularly smoky day, that saw her seek refuge at home, provides insight into the exigent need to counter the weight of climate anxiety with the hopeful risk of planting a seed that may flourish into new life. The seed represents hope in the face of climate anxiety; a chance to plant the seed for an alternative existence.

Within *Resurgence*, the somewhat familiar world allows characters to become emotional surrogates for the actors by cradling the difficult emotions (i.e. climate anxiety) that are evoked by the lived intensities of the climate crisis. Granting the characters to become emotional surrogates empowers the actors (students) to process their eco-anxiety through the choices and consequences that are part of the world building. Speculative fiction, specifically solarpunk, allows for teachers and students to bring a mirror to the contemporary world and explore paths through and beyond the current realities that threaten our existence by imagining an existence that is not restrained by the world we know. Another teacher participant from the research project articulates that,

> It's like a kind of rehabilitation is taking place. After this project is done, I will retain the knowledge that, as a creative community, my students and I can face the sadness, anger, frustration, and fear brought about by the climate crisis and get on with the urgent work that needs to be done, in our classroom and the world beyond.

The emotional composting that is inherent to drama-based solarpunk pedagogies allows for climate anxiety, manifested in sadness, anger, frustration, and fear, to be understood from varying perspectives and angles, generating recognition that these emotions are not one dimensional but are layered with complexities, ambiguities, and partiality. Solarpunk worlds are shaped by the need for other ways of knowing and being that resist the oppressive constraints of systems shackled to anthropocentric thinking and capitalist pursuits, while giving life to action that will generate radical alternatives.

Conclusion

Engaging with the ideas of solarpunk to frame the pedagogical choices of arts-based approaches for teaching and learning, such as dramaturgy, creates and makes space to contemplate the realities of the world through imagined futures of living with the world. Solarpunk generates imaginative spaces that encourage teachers and students to encounter the uncertainties that are inherent to the indeterminacy of the future while contending with the ongoing weight of climate anxiety. Working with the counternarratives (Agee, 2023) of solarpunk encourages engagement with future thinking that is directed towards possibility, sustainability, inclusivity, alterity, resistance, and disengagement from present ways of knowing and being. Space is made for students to recognize their agency in seeking out other futures that are not tethered to patriarchal, colonialist, imperialistic, or capitalistic ideologies. The fantasy of utopian futures and the tragedy of dystopian worlds provide escapist speculations that avoid confronting and enacting other ways of living with the world. Solarpunk pedagogies provide space to explore how we might embrace climate anxiety, turn towards the present-day climate emergencies, acknowledge the historical events that contributed to the contemporary planetary crises, and envision a future that dares to confront the trouble we are in to live in different ways that might mitigate the destruction of our existence. As Beattie remarks about teaching in a world that continues without the guarantee of a future, "This is beautiful, emotional, and risky work" (line 103). Teaching and learning in a time rife with environmental crises necessitate work that disrupts the reproductive tendencies of education systems. We are challenged to re/configure other ways of knowing and being in a world where the future is not certain nor a guarantee of solace from the embodied emotional pressures of climate anxiety.

References

Adams, C. J., & Gruen, L. (Eds.). (2022). *Ecofeminism: Feminist intersections with other animals and the earth* (2nd ed.). Bloomsbury.

Agee, S. (2023). A future dream: How solarpunk helped alleviate my existential dread. *Earth Island Journal, Spring Issue*, 25–32. www.earthisland.org/journal/index.php/magazine/entry/solarpunk-imagines-future-renewable-tech-socio-ecological-enlightenment/##

Ahmed, S. (2010). *The promise of happiness*. Duke University Press.

Apple, M. W. (2018). Critical curriculum studies and the concrete problems of curriculum policy and practice. *Journal of Curriculum Studies, 50*(6), 685–690. https://doi.org/10.1080/00220272.2018.1537373

Association of Canadian Deans of Education. (2022). *Accord on education for a sustainable future*. www.ACDEAccords.ca

Bentz, J. (2020). Learning about climate change in, with and through art. *Climate Change, 162*, 1595–1612. https://doi.org/10.1007/s10584-020-02804-4

Breault, D. A. (2010). Pedagogy. In C. Kridel (Ed.), *Encyclopedia of curriculum studies* (p. 635). Sage.

Bryan, A. (2020). Affective pedagogies: Foregrounding emotion in climate change education. *Policy & Practice: A Developmental Education Review, 30*, 8–30. www.develop menteducationreview.com/issue/issue-30/affective-pedagogies-foregrounding-emotion-climate-change-education#:~:text=The%20Affective%20Pedagogies%20framework%20is,which%20we%20are%20embedded%20and

Climate Psychology Alliance. (2022). *Handbook of climate psychology*. www.climatepsy chologyalliance.org/index.php/component/content/article/climate-psychology-handbook?catid=15&Itemid=101

Davenport, L. (2017). *Emotional resiliency in the era of climate change: A clinician's guide*. Jessica Kingsley Publishers.

Dědinová, T., Łaszkiewicz, W., & Borowska-Szerszun, S. (2021). *Images of the Anthropocene in speculative fiction: Narrating the future*. Lexington Books.

Farrell, A. J. (2022a). *Ecosophy and educational research for the anthropocene: Rethinking research through relational psychoanalytic approaches*. Routledge. https://doi.org/10.4324/9781003024873

Farrell, A. J. (2022b). To love and to teach other people's children in the face of the climate crisis. In A. J. Farrell, C. Skyhar & M. Lam (Eds.), *Teaching in the anthropocene: Education in the face of environmental crisis* (pp. 85–95). Canadian Scholars | Women's Press.

Farrell, A. J. (2023). Animality-as-curriculum for the anthropocene. *Journal of the American Association for the Advancement of Curriculum Studies, 15*(2). https://ojs.library.ubc.ca/index.php/jaaacs/issue/current

Flynn, A. (2014). Solarpunk: Notes toward a manifesto. *Hieroglyph*. https://hieroglyph.asu.edu/2014/09/solarpunk-notes-toward-a-manifesto/

Galway, L. P., & Field, E. (2023). Climate emotions and anxiety among young people in Canada: A national survey and call to action. *The Journal of Climate Change and Health, 9*, 100204. https://doi.org/10.1016/j.joclim.2023.100204

Haraway, D. J. (2016). *Staying with the trouble: Making kin in the chthulucene*. Duke University Press.

Intergovernmental Panel on Climate Change. (2022). *Climate change 2022: Impacts, adaptation and vulnerability: Summary for policymakers* [Contribution of working group II to the sixth assessment report of the intergovernmental panel on climate change]. Cambridge University Press.

Klein, N. (2014). *This changes everything: Capitalism vs the climate*. Vintage Canada.

Learning for a Sustainable Future. (2022). *Canadian perspectives on climate change & education*. https://lsf-lst.ca/wp-content/uploads/2023/03/Canadians-Perspectives-on-Climate-Change-and-Education-2022-s.pdf

Orange, D. M. (2017). *Climate crisis, psychoanalysis, and radical ethics*. Routledge. https://doi.org/10.4324/9781315647906

Pihkala, P. (2018). Eco-anxiety, tragedy, and hope: Psychological spiritual dimensions of climate change. *Zygon, 52*(2), 545–569. http://doi.org/10.1111/zygo.12407

Pinar, W. F. (2019). *What is curriculum theory?* (3rd ed.). Routledge.

Pitt, A., & Britzman, D. (2003). Speculations on qualities of difficult knowledge in teaching and learning: An experiment in psychoanalytic research, *International Journal of Qualitative Studies in Education, 16*(6), 755–776. https://doi.org/10.1080/095183903 10001632135

Rousell, D., & Cutter-MacKenzie-Knowles, A. (2020). A systematic review of climate change education: Giving children and young people a "voice" and a "hand" in redressing climate change. *Children's Geographies, 18*(2), 191–208. http://doi.org/10.1080/1 4733285.2019.1614532

Ruitenberg, C. W., & Rathje, E. (2022). Perceiving the limits, or: What a pandemic has shown us about the climate crisis. *Philosophical Inquiry in Education, 29*(1), 72–77. https://doi.org/10.7202/1088385ar

Tanner, D. (2021). The development of realist speculative narratives to represent and confront the anthropocene. In T. Dědinová, W. Łaszkiewicz & S. Borowska-Szerszun (Eds.), *Images of the anthropocene in speculative fiction: Narrating the future* (pp. 235–255). Lexington Books.

Thornton, S., Graham, M., & Burgh, G. (2019). Reflecting on place: Environmental education as decolonisation. *Australian Journal of Environmental Education, 35*(3), 239–249. https://doi.org/10.1017/aee.2019.31

Tian, M. (2023). *Arts-based research methods for educational researchers*. Routledge. https://doi.org/10.4324/9781003196105

Tomin, B. (2020). Worlds in the making: World building, hope, and collaborative uncertainty. *Journal of the American Association for the Advancement of Curriculum Studies, 14*(1).

Trott, C. D., Even, T. L., & Frame, S. M. (2020). Merging the arts and sciences for collaborative sustainability action: A methodological framework. *Sustainability Science, 15*, 1067–1085. https://doi.org/10.1007/s11625-020-00798-7

Verlie, B. (2021). *Learning to live with climate change: From anxiety to transformation.* Routledge. https://doi.org/10.4324/9780367441265

Wagner, P., & Wieland, B. C. (2022). Introduction: The situation so far. In P. Wagner & B. C. Wieland (Eds.), *Almanac for the anthropocene: A compendium of solarpunk futures* (pp. 1–13). West Virginia University Press.

Weintrobe, S. (2013). The difficult problem of anxiety in thinking about climate change. In S. Weintrobe (Ed.), *Engaging with climate change: Psychoanalytic and interdisciplinary perspectives* (pp. 33–47). Routledge.

Weintrobe, S. (2021). *Psychological roots of the climate crisis: Neoliberal exceptionalism and the culture of uncare.* Bloomsbury.

Więckowska, K. L. (2022). Appositions: The future in Solarpunk and post-apocalyptic fiction. *Text Matters (Łódź), 12*, 345–359. https://doi.org/10.18778/2083-2931.12.21

Woynarski, L. (2020). *Ecodramaturgies: Theatre, performance, and climate change.* Palgrave Macmillan.

9 Feeling Climate Change – Supporting Students Mental Well-being Through Theatre of the Oppressed

Josefina Rueter Veiga

Positionality Statement

Over years of teaching and researching climate change education, I have come to understand that it is a multi-dimensional endeavour that requires all of us to go beyond just knowing the facts. My experience teaching in Iqaluit, Nunavut, an area experiencing significant changes due to climate change, opened my perception of the magnitude of the impacts of climate change and the multiple ways of knowing and understanding the topic. Coming from a science education background, considering the affective component of climate change was an important shift for me. After living in Nunavut for ten years, and subsequent climate education and activism involvement in Ontario, I feel strongly that today's youth deserve educational opportunities on climate change that will support and prepare them for the challenges they will face in the future.

> *Perhaps the theatre is not revolutionary in itself; but have no doubts it is a rehearsal for the revolution.*
>
> (Boal, 1974/2008, p. 134)

The Intergovernmental Panel on Climate Change has researched and documented the causes, impacts, and progress of climate change over the past two decades. It has been clearly documented that global climate change is due to anthropogenic causes and its effects are widespread and fraught with inequity. Unprecedented environmental challenges, such as extreme weather events, are disproportionately affecting those who have contributed least to global emissions. This will continue to be the case. One such group is children and youth. The risk to children and youth's physical health has been well documented and includes heat-related illnesses, environmental toxin exposures, and infectious and gastrointestinal diseases (Sanson et al., 2019). More recently, the impacts of climate change on mental health have been researched and include increases in post-traumatic stress disorder (PTSD), depression and anxiety, and sleep issues (Clayton, 2020; Sanson et al., 2019). Whether through direct experiences with

DOI: 10.4324/9781003494416-11

climate change impacts or indirect experiences, such as media exposure, students are experiencing emotional responses to climate change (Ojala et al., 2021). As such, it is essential for students to have supportive environments, including in schools, in which to explore, share, and process their emotions (Lehtonen & Pihkala, 2021).

Unfortunately, as Irwin (2020) states "at present, the climate crisis has hardly been noticed by the educational establishment, even if school students are now leading the challenge by going on climate change strikes. The overt curriculum in most areas is unchanged by climate awareness" (pg. 493). Where climate change does enter the formal educational sphere, science-based approaches are most common (Monroe et al., 2019, Rousell & Cutter-Mackenzie-Knowles, 2020; Trott, 2020). The curriculum rarely engages with the "intersecting social, political, cultural, and economic components of climate issues, or for building the skills needed to participate in collective planning to address uncertain futures" (Karsgaard & Davidson, 2023, p. 77). As a result, different approaches to climate change education that focus on emotional and participatory responses are needed (Cutter-Mackenzie & Rousell, 2019). Lehtonen and Pihkala (2021) argue for "a need for collective and creative approaches, such as performing-making, that can address the complex psychosocial issues and young people's ideas of climate change" (p. 757). Baker et al. (2021) stress the important responsibility that schools have to make sure students are learning about climate change and supporting the emotional well-being during their learning.

Historically, the arts have been used as a method for promoting social change, communicating injustices, and engaging populations in a variety of issues. More recently, arts-based methods are gaining recognition as effective and engaging methods for climate change education (Bentz & O'Brien, 2019). Trott et al. (2020) specifically argue for the need for art-science integration "to activate the transformative potential of the arts for social change" (p. 1068). The arts not only offer transformative potential but also engage learners in a multitude of ways, creating emotional and personal connections (Bentz, 2020). Art can also be healing (Marshall, 2014) and persuasive (Silva & Menezes, 2016).

This chapter explores an approach to climate change education that focuses on the emotional, creative, and participatory (Cutter-Mackenzie & Rousell, 2019) responses by utilizing a theatre-based approach. Based on Theatre of the Oppressed (TO), students participated in various image theatre exercises and forum theatre plays to explore and engage with information about and issues related to climate change. The original study examined the efficacy of this approach with respect to climate change education generally, the impact on students' climate change literacy and agency on taking action, and how it could help students explore and process their feelings and emotions around climate change. However, this chapter will focus mainly on how students explored their feelings and emotions around climate change. It will additionally consider how learning about climate change and developing agency supports students' mental well-being with respect to climate change.

About Climate Drama

The study took place in a large urban school board in South-Central Ontario, Canada. The students ranged in age from 8 to 13 years old, were from diverse cultural and socio-economic backgrounds, and included students with diverse learning needs. While the study approached climate change from a mostly Global North, industrialized context, the theatre approach used (Theatre of the Oppressed) has been used widely across the globe in a variety of contexts with diverse participants (Crean, 2013; Davidson, 2021; Rhoades, 2021; Schutzman & Cohen-Cruz, 2002; Ventä-Olkkonen et al., 2022). While the students who participated in the study have not necessarily experienced the more severe impacts of climate change[1] (extreme weather, drought, floods, etc.), it has been noted that indirect experiences of climate change through media and education can also lead to anxiety and other emotional reactions (Clayton, 2020; Lehtonen & Pihkala, 2021; Ojala et al., 2021; Sanson et al., 2019).

Climate Drama was collaboratively outlined over three focus group meetings by four elementary school teachers, an environmental scientist, and the researcher. Student input was incorporated throughout programme implementation (from pre-programme interviews, student feedback during sessions) to ensure student interests and voices were included. Given the absence of climate change curricular outcomes at the elementary level, the programme was designed to meet Drama, Environmental Education, and at times, Science curricular outcomes. The participating teachers incorporated the programme as part of their Drama curriculum. Climate Drama consisted of six, approximately one-hour sessions that took place over the span of six to eight weeks during the regular school day. The sessions were led primarily by the classroom teacher with support as requested. In total, four elementary classes participated (four teachers and approximately 65 students) with students in Grades 3 to 7. The four teachers included one Grade 3 teacher, one Grade 3/4 teacher, one Grade 5/6 teacher, and one Grade 7 drama/dance teacher. Data collection included pre- and post-programme interviews with students and teachers, observations and field notes, photographs, and video recordings.

Sessions[2] were structured as follows: students listened to a reading of a storybook related to climate change, participated in a variety Theatre of the Oppressed (TO) activities (including Image Theatre and Forum Theatre), and reflected on their learning. Themes and topics covered included climate change basics (definition, causes, greenhouse effect and gases, impacts), what students love and want to protect about the Earth, emotions related to climate change, how to be in good relationship to the Earth, and actions that can be taken. In the final two sessions, students created and performed their own

Forum Theatre plays. For these plays, students presented an issue related to climate change (such as food waste) to their classmates; the classmates as spect-actors[3] stepped into the plays and suggested alternative actions to the issues to support positive change.

All four participating teachers felt comfortable supporting their students through challenging emotions. Only one participating teacher felt that this might be a concern for their students; however, this teacher had been working with the same class for several years and had a strong relationship with the students. All teachers were provided with some grounding exercises to use with students in case students were feeling overwhelmed or struggling with their emotions. Additionally, the sessions avoided engaging in a doom and gloom narrative around climate change and instead focused on engaging students on actions and changes they could make.

Theatre of the Oppressed

The theatre-based climate change programme was based upon Augusto Boal's Theatre of the Oppressed (TO) (1974/2008) whose purpose is for "people to analyze oppressions, share knowledge with one another, and mobilize the knowledge they already have. Ideally, they collectively mobilize that knowledge to dismantle oppressive systems" (Howe et al., 2019, p. 1). Performed in non-traditional settings, TO engages participants in participatory, embodied, and improvizational exercises and techniques to teach basic theatre skills as well as help participants understand how using their bodies in theatre can support change in their own lives (Boal, 1974/2008; Erel et al., 2017; Marín, 2007). This research study used two TO techniques: Image Theatre and Forum Theatre. In Image Theatre, participants create images to represent an idea, concept, moment, or feeling. These images are viewed and explored, and at times altered, encouraging everyone to engage in whichever way resonates most deeply. Within Image Theatre, Boal (1992/2022) developed numerous exercises, with multiple variations, focusing on embodying participant experiences. During Forum Theatre, participants collaboratively develop and perform a short play on a social issue or oppression that is relevant to their own lives. The play presents a clear and unresolved conflict, which spect-actors attempt to resolve. Boal's work has been "described as a system, open to reinterpretation and adaptation to a wide range of contexts and a bewildering plurality of content" (Campbell, 2019, p. 6). This can be seen in Image Theatre and Forum Theatre as both techniques focus on eliciting participants' lived experiences and knowledge. As such, the programme itself, and its open and flexible approach, would be transferable to other contexts (Babbage, 2004).

In considering the roles youth play in climate change discourse, and even in society more broadly, Boal (2010) states,

Theatre is the best place in which anyone can experiment with what it is to be someone else . . . If it is society that constructed those roles, then we can imagine a society in which those roles will be different.

(p. xvi)

Youth are, through climate marches and other forms of activism, becoming a larger part of the discussion on climate change. TO, particularly Forum Theatre, can provide opportunities for youth to practice for these roles and reimagine different futures. These experiences can be incredibly empowering for students and encourage them to speak more openly about climate change (Davidson, 2015).

TO, as a participatory process, can increase trust and help develop a greater sense of capability and control (Ojala et al., 2021). This is particularly important for topics such as climate change, especially to help combat feelings of anxiety and helplessness. Image Theatre and Forum Theatre provide a way to engage emotionally with climate change and connect with others on issues such as anxiety and worry (Lehtonen & Pihkala, 2021). As Marín (2007) found in her research, "theatre offered the participants a safe environment in which to express themselves and examine how they felt about [the] issues" (p. 92). These aspects allow for a different type of engagement with climate change learning and action.

Pre-Programme Emotions and Worries

Through the pre-programme interviews, students were asked about their feelings and worries about climate change. Most students felt negatively about climate change – students expressed feeling worried, scared, sad, angry, and just not liking climate change. One student stated: "Honestly, pretty scared, because it's definitely going to affect us in our lifetime . . . more so than it is now." A few students spoke about feeling sad for animals, "it makes me worried for all those animals and nature that is getting ruined," and "like also sad because there's a whole bunch of animals that like won't be around, like when I'm gonna be an adult." While others expressed wanting something to be done, "But I feel like a lot of people say like they want to do something about it. They want to fix it, but no one actually ever does anything." There were a few comments about feeling ok about climate change and that there could potentially be good things that come from it.

Students' worries mostly centred on the impacts of climate change, potential environmental changes, and worsening weather events (storms, floods, and increasing heat). Specifically, students focused on the impacts on animals (including habitat loss), "I worry that a lot of animals or like plants won't learn to adapt because it's getting too hot for them to live, and they'll go extinct." Some students showed concern for how climate change will affect humans – human extinction and the end of the world. Others were worried about how inaction will place the burden of climate change on their generation, "That in, like, the future that we're gonna have to like, like the kids our age are gonna have to like, step up and help a lot with the environment." None of these results is particularly surprising given the plethora of research that has already documented how children and youth are feeling about climate change.

Exploring Emotions Through Image Theatre

Image Theatre is a technique found in Theatre of the Oppressed in which participants create stationary images or scenes that represent an idea, emotion, or theme. Participants are asked to "engage with their own experiences very directly" (Babbage, 2004, p. 104) when silently posing themselves or others. When writing about Image Theatre, Boal (1992/2022) states "we should not try to 'understand' the meaning of each image . . . but to feel those images, to let our memories and imaginations wander" (p. 175). It is clear that feeling and emotional embodiment are at the heart of Image Theatre. Image Theatre allows participants to connect to a topic without having to use words. This is especially helpful for emotionally laden topics, such as climate change, as students may not have the words or capacity to adequately voice how they are feeling. Engaging with emotions in the school setting has been shown to help students process how they are feeling as well as decreasing the chances of feeling isolated, stuck, anxious, or sad (Lehtonen & Pihkala, 2021).

Throughout the first four sessions, students participated in various Image Theatre exercises. As an introduction to Image Theatre, students were given single words, such as heat, sun, Earth, to practice creating visual representations and holding the poses quietly. Students then worked on creating more complex images using other students or as a group. By the end of the programme, students had created images of climate change, greenhouse effect, impacts of climate change, solutions to climate change, emotions on climate change, and actions that could be taken on climate change. The activities described later took place during various sessions and are not presented in chronological order.

Love

The first emotion the students actively examined through the programme was that of love. It was important that the students explored what they love, what they love about the environment, what about the environment is important to them, and subsequently what they would want to protect. When students can identify with and relate to things that are important to them (and their community, family, friends), there is a greater chance they will be willing to take action on climate change.

Using the Image Theatre technique Image of the Word: illustrating a subject with your body (Boal, 1992/2022), students created individual poses of what they love, something they love about the environment, or something about the environment that is important to them. All three categories were reflected in the images students created:

- Kicking a ball, skipping, baseball, hockey, horseback riding;
- Ants, trees, snow, nature; and
- Looking at the stars, climbing trees, skiing, reading a book in nature.

Figure 9.1 shows students expressing what they love or what they love about the environment by posing as a tree, going horseback riding, and playing hockey.

Figure 9.1 Image Theatre Example – What We Love

With love as a starting point to connect to the environment, students were then asked to consider what elements of Earth or the Environment they would want to protect. Students participated in both self-posed images and group sculpts and created images of:

- The Earth, trees, sun, bush, flowers, mountains, plants;
- Human protecting themself, a kind person, turning off the lights, stopping cars, kids; and
- Army ant, panda, polar bear, reindeer, birds, bunny, dogs, insects.

Figure 9.2 shows students sharing poses of ways to protect the environment (turning off the lights, stopping cars) and what they want to protect (plants, trees, and bushes).

By focusing on positive emotions related to the environment, students engage in meaning-focused coping (Ojala, 2012) – they consider what is important to them, what they value, and how to go about protecting what is important.

Figure 9.2 Image Theatre Example – What We Want to Protect

I Feel...

Students also participated in an activity called The Introspective Techniques: Image of the Antagonist (Boal, 1995/2013) to explore and share their feelings about climate change. This technique engages each individual student in thinking about their own individual emotional relationship with an "antagonist" – in this case climate change. To begin, students stood in a circle facing outwards. They were asked to think about how they felt about climate change and how they could show that feeling in a stationary pose. Once all students were ready, they turned to the inside of the circle and entered their poses. Students were then asked to silently gather with other students who seemed to be showing the same emotion. Once grouped together, each group silently selected one student to represent their whole group. These individuals then moved to the front of the classroom and shared what emotion or how they were feeling about climate change.

Figure 9.3 Image Theatre Example – How We Feel About Climate Change

Much like the pre-programme interviews, students expressed negative emotions about climate change:

- "Kinda mad that people are taking away . . . climate change is taking away animal's food."
- "I was sad because people are making the smoke . . . doing what they are doing but not stopping."
- "I was really tired."
- "I was thinking why do people have to be doing this?"
- "Why, why, I don't want to do this. I am sad."

The group members were then given the opportunity to share if they felt the same as the individual who shared – for the most part everyone expressed the same feeling as the individual in their group who shared. The students in Figure 9.3 all shared they were feeling upset about climate change and mad that climate change was taking away food from animals. Students who had difficulty expressing their emotions verbally were able to focus on the embodiment portion of the activity rather than opt out (Pihkala, 2020). It was wonderful to see how students could share their emotions with and without words and were able to interpret others' poses. Both non-verbal and verbal sharing of emotions are important in helping students process their feelings around climate change (Ojala et al., 2021).

What Is Climate Change?

Throughout the programme, students participated in various Image Theatre exercises around climate change and related concepts, such as the greenhouse effect. The instructions given to the students were purposefully vague (create an image of what you think of when you hear the words climate change) so that students could choose how to approach the topic. While these were intended to help students develop a greater understanding of climate change (for example, causes and effects), at times, students chose to depict emotional reactions and feelings instead. This clearly shows the emotional impact climate change can have on students.

In one activity using the technique Image of the Word – illustrating subjects using other people's bodies (Boal, 1992/2022) – the image creators placed three students in a semi-circle with their hands up and fingers splayed open and three other students with various expressions (worry, fear). When asked to explain their image, the creators shared that the three students in the semi-circle were the sun and the other three were worried about climate change. It was clear that to them, climate change meant worry and fear. In another activity called Group Sculpt (Boal, 1992/2022), students first posed individually in response to the topic of climate change. They were then asked to move together into a group in relation to each other to create an image of climate change. In the middle of this image, two students came together hugging and crying. Again, their response to climate change was emotional rather than a cause or impact. These instances show that the students were able to use the drama exercise to help explore and engage with their emotions, which is helpful when dealing with worry about climate change (Ojala et al., 2021).

While some students expressed worry and sadness, some students were able to use humour, a strategy shown to increase resilience (Pihkala, 2020), when thinking about climate change. Figure 9.4, another Image of the Word illustrates subjects using other people's bodies (Boal, 2022), a student was asked to create an image of climate change. He placed three students in a row. He shared that they were the Earth, a tree, and a "muscle man." This last naming elicited lots of laughing and giggling. The exchange between the teacher and student regarding "muscle man" was very enlightening:

Teacher: "Why did you chose muscle man?"
Student: "To fight off the greenhouse gases that are bad."
Teacher: "Why do you need a muscle man to fight off greenhouse gases?"
Student: "To have a better Earth."
Teacher: "Why did you chose a muscle man?"
Student: "Because he's strong."

The student was not only able to use humour to respond to climate change but also showed a great deal of understanding about the role of greenhouse gases and responsibility to take action.

Figure 9.4 Image Theatre Example – What Is Climate Change

Exploring How Non-Humans Feel

Empathy for nature and the more-than-human inhabitants has been found to positively influence pro-environmental attitudes and behaviours (Ienna et al., 2022). To help students build empathy for the more-than-human, they participated in an activity imagining how animals or plants/trees would be feeling in different situations (Westbury & Neumann, 2008). Using a technique called Animals or Vegetables in Emotional Situations (Boal, 1992/2022), students selected an animal, plant, or tree to embody. They were then given different situations related to climate change such as intense storms, habitat destruction, drought, extreme weather (heat), and human encroachment on habitats. Students were instructed to react as if they were that animal, plant, or tree and were free to use both movement and sound. A few examples of what students enacted include

- " panda trying to hide and running away from the storm";
- A tree trying to keep its roots in soil;
- Animals trying to protect their babies; and
- A penguin on a glacier being flooded.

It was clear the students were very engaged in imagining events from their chosen more-than-human perspective. In addition to developing empathy for the more-than-human, this activity provided students with the chance to consider alternative perspectives to the climate crisis – not just how it affects humans but the larger world around them.

Supporting Emotions Through Knowledge and Action

Understanding Climate Change

Climate anxiety can be exacerbated by a lack of knowledge as this causes increased uncertainty and confusion (Crandon et al., 2022). It was clear from the focus group sessions and the interviews with teachers and students that the students needed more information about climate change to be able to fully participate in the drama programme. A significant portion of the students interviewed (36.6%) stated that they did not know what climate change was or they had forgotten. Students were most knowledgeable about the effects and impacts of climate change (such as temperature changes and environmental changes). Few knew the causes of climate change and multiple misconceptions surfaced, including confusing weather with climate, the seasons, or climate zones. Given that most of the student participants were Grade 3 and 4 students, it is understandable that the majority of participants did not know a lot about climate change. As a result, it was determined that there was a need for some instruction, both direct and through storybooks, with a focus on defining climate change, the greenhouse effect and greenhouse gases, and the main causes of climate change. Any instruction was always followed up with TO activities. Helping students grow a deeper and accurate understanding and awareness of climate change in intentional and supportive ways can help students better cope with emotional responses to climate change (Baker et al., 2021; Crandon et al., 2022).

In the post-programme interviews, on the question of "What do you know now about climate change?" almost all students shared new information they had learned. Only six stated they didn't learn anything – however all students were able to share information on actions that could be taken to help address climate change.

Rehearsing for the Climate Revolution

During the pre-programme interviews, students indicated a clear interest in learning how to stop climate change and what they could do. The teachers also emphasized a desire to focus the programme on solutions and actions students could take to create change and make a difference.

Exploring actions and solutions to climate change can help students manage their emotional responses (Baker et al., 2021; Pihkala, 2020). When students participate in experiences that lead to a sense of increased agency, they also feel a greater sense of hope about the future and are more likely to engage in action (Crandon et al., 2022; Li & Monroe, 2019; Stevenson & Peterson, 2015).

Through Image Theatre students created images of how to be in good relationship to the Earth and any action they felt would make a positive difference with respect to climate change. Most actions that students depicted were individual-based actions such as caring for plants, animals, and trees; planting trees and plants; reading a book about climate change; picking up garbage; and not using cars. A few students showed more collective participatory actions such as protesting and teaching with signs.

During the final two sessions of the drama programme, students created and delivered Forum Theatre plays on a variety of issues related to climate change. Forum Theatre (Boal, 1974/2008, 1992/2022) is a form of participatory, interactive theatre that engages the audience members, known as spect-actors, in addressing a conflict (usually a form of oppression or social injustice). The Forum Theatre plays provided students with the opportunity to "grapple with an issue, event, or question of immediate public and personal concern" (Taylor, 2003, p. 7).

Forum Theatre creates opportunities for all participants to improvise and "realign themselves in relation to the changing action" (Babbage, 2004, p. 45) which provides multiple opportunities to try different climate interventions. Boal argues that one purpose of Forum Theatre is to create energy and encourage the carrying over of that energy into real life (Babbage, 2004). For the spect-actors who intervene, "they see *themselves* in the act of intervention and know themselves to have the power to transform their capacity for resistance, transformation and revolution into lived, social reality" (Campbell, 2019, p. 16). In small groups, the students first developed a short play ending with a protagonist facing an unsolved problem related to climate change. Each group presented their play once without audience interaction. Subsequently, each group presented their play with spect-actors stopping the play, entering the scene to replace characters, and trying out different actions to solve the problem.

Overwhelmingly, the students shared that the Forum Theatre plays were their favourite part of the programme. Students enjoyed both acting out their group-created scenarios and being spect-actors who intervened. Even those who did not actively intervene stated they enjoyed the plays. While not all interventions were successful in the Forum Theatre context, students still expressed excitement about being able to engage in trying to change the outcome. In one example, a group presented a forum theatre play where the father only ate red meat. The family attempted to convince the father to eat less red meat. Spect-actors stepped in to attempt to change the father's mind. Their arguments included being healthier, living longer, saving money, and meat-free alternatives. All were met with a resounding no by the father, until one student offered to do the cooking!

Post-Programme Emotions

The post-programme student interviews did not show any major change in how the students were feeling about climate change. The overall results indicated that students remained feeling negatively, angry, scared, frustrated about inaction, and sad.

- "Kind of scared about it. Because, I'm not saying you in particular, but adults are leaving a messed-up world for our generation."
- "Scared . . . because we're pretty deep into it. And show our generation like we're the ones that are supposed to fix it."
- "Now I'm more upset because climate change is ruining lives of poor innocent animals and changing and ruining our environment. I'm very mad at it. Very very mad."

Interestingly, several students, even while feeling negatively about climate change, shared that they were happy to learn more about it – "Sad, but happy that I learned about it." Similarly, the students were not feeling very hopeful or good about the future of the environment, including some students feeling worse now that they knew more about climate change. However, some students did indicate that they would feel better and more hopeful if action is taken soon.

While at first this may seem to be a failure of the programme, feeling negatively about climate change is not necessarily something to fix. In a review of literature on anxiety, worry, and grief related to climate change, Ojala et al.

(2021) found that negative views, concerns, and worry can lead to more pro-environmental behaviour and a greater agency to act on climate change; these are all rational responses to a threat like climate change (Pihkala, 2020). This was demonstrated in the post-programme interviews, as every single student interviewed was able to identify at least one action that they could take to make a positive change related to climate change.

Benefits

Throughout the programme, the students engaged with an emotionally and politically charged topic in different ways that provided them with creative outputs open to a variety of interpretations and opportunities to express a wide range of emotions. This flexibility and openness enabled everyone to find a way to engage with climate change at a level they felt comfortable with. Image Theatre and Forum Theatre provide a way to engage emotionally with climate change and connect with others on issues such as anxiety and worry (Lehtonen & Pihkala, 2021).

Observations and review of interview data show evidence of increased knowledge and understanding of climate change and increased agency. This is important, as Crandon et al. (2022) indicate, a "[l]ack of information about how to mitigate climate change also limits a young person's ability to manage anxiety using action" (p. 127). Additionally, Ojala (2012) found that problem-focused coping (a strategy used in this programme) helps to regulate worry.

Through the Forum Theatre productions, students experienced the opportunity to build their "sense of self-efficacy" (Sanson et al., 2019, p. 203), helping combat the feeling of powerlessness.

The programme also provided students with a place and opportunity to safely explore climate change, its causes and impacts, thereby allowing students to share their own feelings – with and without words (Crandon et al., 2022, p. 127). This form of emotion-focused coping was also positively correlated to regulating worry (Ojala, 2012). While students did express worry and anxiety about the future and climate change in the post-programme interviews, it seems that rather than falling into hopelessness, the students were able to use those feelings to develop steps to action (Crandon et al., 2022; Ojala, 2012; Verplanken & Roy, 2013).

While not directly part of the data collected, several anecdotal instances merit mentioning. One student created "The Book of Climate Change" with information about climate change, including what students could do to help and a quiz. Two parents spoke to me about how their children were coming home and talking about the activities we had done in class and wanting to do more. One child even had his family create an image demonstrating the greenhouse effect. Lastly, during one of the post-programme teacher interviews, the teacher mentioned the students speaking more about climate change and some selecting to use it as the topic for their persuasive writing assignment. These small snippets show how students transfer their learning and engage with it in multiple new ways.

Future Considerations

Theatre of the Oppressed has been well studied in various disciplines but is only beginning to be researched with respect to climate change. Future studies using TO to engage and teach about climate change could consider follow-up interviews at longer post-programme intervals to examine how participants were able to transfer their knowledge, agency, and empowerment into their own lives, as well as how students are coping with climate anxiety and other emotions. A further element that could potentially be added to future programmes using TO is to have students engage in a real-life action after the conclusion of the drama programme and then investigate how TO supported their engagement and turned rehearsal into revolution. TO provides multiple entry points for educators wishing to utilize TO with their students to explore issues around climate change. While this study focused on Image Theatre and Forum Theatre, other techniques within TO, such as Legislative Theatre, Newspaper Theatre, and Invisible Theatre, are alternative options that educators could explore.

Conclusion

This research study explored the various ways in which students can explore their emotions around climate change in a safe and supportive manner. Climate change continues to impact the natural world and the mental and emotional well-being of children and youth; therefore, it is important to encompass the affective dimension in climate change education experiences. Bentz (2020) expresses the need for more research to be done with regard to teaching about climate change "in, with, and through the arts" (p. 1597) and Rousell and Cutter-Mackenzie-Knowles (2020) emphasize the need for more creative and participatory approaches to climate change education, moving beyond purely knowledge-based approaches and research studies. With a focus on embodiment and engagement, Boal's approach can address emotions and help students move past a sense of overwhelm (Davidson, 2021). As such, Theatre of the Oppressed shows great potential as a method for learning and engaging with climate change.

Notes

1 One Grade 3 student spoke about their experiences with the wildfires in Kelowna, BC.
2 The 6th session was the exception to this structure. During the 6th session, the entire time was dedicated to the Forum Theatre plays.
3 Spect-actors are spectators (audience members) turned actors through their participation in a Forum Theatre play (Boal, 1974/2008).

References

Babbage, F. (2004). *Augusto Boal*. Routledge.

Baker, C., Clayton, S., & Bragg, E. (2021). Educating for resilience: Parent and teacher perceptions of children's emotional needs in response to climate change. *Environmental Education Research, 27*(5), 687–705. https://doi.org/10.1080/13504622.2020.1828288

Bentz, J. (2020). Learning about climate change in, with and through art. *Climatic Change, 162*(3), 1595–1612. https://doi.org/10.1007/s10584-020-02804-4

Bentz, J., & O'Brien, K. (2019). Art for change: Transformative learning and youth empowerment in a changing climate. *Elementa: Science of the Anthropocene, 7*(52), 1–19. https://doi.org/10.1525/elementa.390

Boal, A. (2008). *Theatre of the oppressed* (C. A., M.-O. L. McBride & E. Fryer, Trans.). Pluto Press (Original work published 1974).

Boal, A. (2013). *The rainbow of desire: The Boal method of theatre and therapy* (A. Jackson, Trans.). Routledge (Original work published in 1995). http://doi.org/10.4324/9780203820230

Boal, A. (2022). *Games for actors and non-actors* (3rd ed., A. Jackson, Trans.). Routledge (Original work published in 1992). http://doi.org/10.4324/9780429261053

Boal, J. (2010). Our role in crisis (R. Cave, Trans.). In P. Duffy & E. Vettraino (Eds.), *Youth and theatre of the oppressed* (pp. xv–xvii). Palgrave MacMillan.

Campbell, A. (2019). *The theatre of the oppressed in practice today: An introduction to the work and principles of Augusto Boal*. Bloomsbury Publishing.

Clayton, S. (2020). Climate anxiety: Psychological responses to climate change. *Journal of Anxiety Disorders, 74*, 1–7. https://doi.org/10.1016/j.janxdis.2020.102263

Crandon, T. J., Scott, J. G., Charlson, F. J., & Thomas, H. J. (2022). A social–ecological perspective on climate anxiety in children and adolescents. *Nature Climate Change, 12*(2), 123–131. https://doi.org/10.1038/s41558-021-01251-y

Crean, M. (2013). Once upon a time in the Bronx: Working with youth to address violence through performance and play. In *Proceedings of the 12th international conference on interaction design and children* (pp. 443–446). https://doi.org/10.1145/2485760.2485806

Cutter-Mackenzie, A., & Rousell, D. (2019). Education for what? Shaping the field of climate change education with children and young people as co-researchers. *Children's Geographies, 17*(1), 90–104. https://doi.org/10.1080/14733285.2018.1467556

Davidson, D. (2021). "Can we talk?" Forum theatre as rehearsal for climate change interventions. In C. Alexandrowicz & D. Fancy (Eds.), *Theatre pedagogy in the era of climate crisis* (pp. 115–129). Routledge.

Davidson, R. (2015). *A stage the earth is going through: Deploying the theatre of the oppressed to address climate change* [Unpublished Bachelor of Arts dissertation, Charles Sturt University].

Erel, U., Reynold, T., & Kaptani, E. (2017). Participatory theatre for transformative social research. *Qualitative Research, 17*(3), 302–312. https://doi.org/10.1177/1468794117696029

Field, E. (2017). Climate change: Imagining, negotiating, and co-creating future(s) with children and youth. *Curriculum Perspectives, 37*(1), 83–89. https://doi.org/10.1007/s41297-017-0013-y

Howe, K., Boal, J., & Soeiro, J. (2019). Introduction. In K. Howe, J. Boal & J. Soeiro (Eds.), *The Routledge companion to theatre of the oppressed* (pp. 1–12). Routledge.

Ienna, M., Rofe, A., Gendi, M., Douglas, H. E., Kelly, M., Hayward, M. W., Callen, A., Klop-Toker, K., Scanlon, R. J., Howell, L. G., & Griffin, A. S. (2022). The relative role of knowledge and empathy in predicting pro-environmental attitudes and behaviour. *Sustainability, 14*(8), 1–21. https://doi.org/10.3390/su14084622

Irwin, R. (2020). Climate change and education. *Educational Philosophy and Theory, 52*(5), 492–507. https://doi.org/10.1080/00131857.2019.1642196

Karsgaard, C., & Davidson, D. (2023). Must we wait for youth to speak out before we listen? International youth perspectives and climate change education. *Educational Review, 75*(1), 74–92. https://doi.org/10.1080/00131911.2021.1905611

Lehtonen, A., & Pihkala, P. (2021). Encounters with climate change and its psychosocial aspects through performance making among young people. *Environmental Education Research, 27*(5), 743–761. https://doi.org/10.1080/13504622.2021.1923663

Li, C. J., & Monroe, M. C. (2019). Exploring the essential psychological factors in fostering hope concerning climate change. *Environmental Education Research, 25*(6), 936–954. https://doi.org/10.1080/13504622.2017.1367916

Marín, C. (2007). A methodology rooted in praxis: Theatre of the oppressed (TO) techniques employed as arts-based educational research methods. *Youth Theatre Journal, 21*(1), 81–93. https://doi.org/10.1080/08929092.2007.10012598

Marshall, L. (2014). Art as peace building. *Art Education, 67*(3), 37–43. https://doi.org/10.1080/00043125.2014.11519272

Monroe, M. C., Plate, R. R., Oxarart, A., Bowers, A., & Chaves, W. A. (2019). Identifying effective climate change education strategies: A systematic review of the research. *Environmental Education Research, 25*(6), 791–812. https://doi.org/10.1080/13504622.2017.1360842

Ojala, M. (2012). Hope and climate change: The importance of hope for environmental engagement among young people. *Environmental Education Research, 18*(5), 625–642. https://doi.org/10.1080/13504622.2011.637157

Ojala, M., Cunsolo, A., Ogunbode, C. A., & Middleton, J. (2021). Anxiety, worry, and grief in a time of environmental and climate crisis: A narrative review. *Annual Review of Environment and Resources, 46*(1), 35–58. https://doi.org/10.1146/annurev-environ-012220-022716

Pihkala, P. (2020). Eco-anxiety and environmental education. *Sustainability, 12*(23), 1–38. https://doi.org/10.3390/su122310149

Rhoades, R. (2021). Nurturing hopeful agency: Applied theatre pedagogy in collaboration with social movements. In C. Alexandrowicz & D. Fancy (Eds.), *Theatre Pedagogy in the era of climate crisis* (pp. 17–31). Routledge.

Rousell, D., & Cutter-Mackenzie-Knowles, A. (2020). A systematic review of climate change education: Giving children and young people a "voice" and a "hand" in redressing climate change. *Children's Geographies, 18*(2), 191–208. https://doi.org/10.1080/14733285.2019.1614532

Sanson, A. V., Van Hoorn, J., & Burke, S. E. (2019). Responding to the impacts of the climate crisis on children and youth. *Child Development Perspectives, 13*(4), 201–207. https://doi.org/10.1111/cdep.12342

Schutzman, M., & Cohen-Cruz, J. (2002). *Playing Boal: Theatre, therapy, activism.* Routledge.

Silva, J. E., & Menezes, I. (2016). Art education for citizenship: Augusto Boal's theater of the oppressed as a method for democratic empowerment. *Journal of Social Science Education, 15*(4), 40–49. https://doi.org/10.4119/UNIBI/jsse-v15-i4-1507

Stevenson, K., & Peterson, N. (2015). Motivating action through fostering climate change hope and concern and avoiding despair among adolescents. *Sustainability, 8*(1), 1–10. https://doi.org/10.3390/su8010006

Taylor, P. (2003, February 20–22). *The Applied Theatre* [Paper presentation]. Arizona State University Department of Theatre Symposium: What is "Cinderella" Hiding?, Tempe, AZ, United States. https://eric.ed.gov/?id=ED479871

Trott, C. D. (2020). Children's constructive climate change engagement: Empowering awareness, agency, and action. *Environmental Education Research, 26*(4), 532–554. https://doi.org/10.1080/13504622.2019.1675594

Trott, C. D., Even, T. L., & Frame, S. M. (2020). Merging the arts and sciences for collaborative sustainability action: A methodological framework. *Sustainability Science, 15*(4), 1067–1085. https://doi.org/10.1007/s11625-020-00798-7

Ventä-Olkkonen, L., Iivari, N., Sharma, S., Juusila-Cevirel, N., Molin-Jusstila, T., Kinnunen, E., Holappa, J., & Hartikainen, H. (2022). All the world is our stage: Empowering children to tackle bullying through theatre of the oppressed in critical design and making. In *Proceedings of the nordic human-computer interaction conference* (pp. 1–15). https://doi.org/10.1145/3546155.3546705

Verplanken, B., & Roy, D. (2013). "My worries are rational, climate change is not": Habitual ecological worrying is an adaptive response. *PLoS ONE, 8*(9), e74708. https://doi.org/10.1371/journal.pone.0074708

Westbury, H. R., & Neumann, D. L. (2008). Empathy-related responses to moving film stimuli depicting human and non-human animal targets in negative circumstances. *Biological Psychology, 78*(1), 66–74. https://doi.org/10.1016/j.biopsycho.2007.12.009

10 Literature and Reading as Ways to Engage With Climate Anxiety in Classrooms

Satu Grünthal, Toni Lahtinen, and Panu Pihkala

Positionality Statement

We are Finnish researchers, living in safe surroundings, in a country which is not (yet) heavily impacted by climate change. We each have special expertise in climate emotions and education: Grünthal is a senior university lecturer in teacher education and a specialist on reading and writing literacy, Pihkala is an interdisciplinary scholar with experience in education, and Lahtinen is a university lecturer, poet, and literature scholar. We are committed to the importance of alleviating climate change for ethical reasons, and we regard climate emotions, including climate anxiety, as important topics to be engaged with in societies.

The Potential of Texts and Reading in Relation to Climate Emotions

The various potentials of texts and reading in relation to climate emotions have been subjects of increasing interest among educators and researchers (e.g. Atkinson & Ray, 2024; Bladow & Ladino, 2018; Turp, 2023); texts and reading can evoke many kinds of climate emotions in readers (Helle et al., 2024; Schneider-Mayerson et al., 2023). Various professionals, such as educators, psychologists, and communication specialists, have been interested in ways in which texts and reading could be used constructively to engage with climate anxiety and other climate emotions (e.g. Bowman et al., 2024; Moser, 2015; Pihkala, 2020a). In this chapter, we approach these issues by drawing from literary criticism, environmental education research, and interdisciplinary research on eco-emotions. We use Finnish education as a practical example, and we utilize many Finnish studies on the intersections of climate emotions, texts, and education.

We use the term *eco-emotions* (similar to environmental and ecological emotions) to refer to affective phenomena which are significantly related to ecological issues, and the term *climate emotions* to refer to cases where there is some focus on anthropogenic global warming in these emotions (for theoretical

DOI: 10.4324/9781003494416-12

discussions, see Bladow & Ladino, 2018; Pihkala, 2022). We recognize that the term *climate anxiety* can be used in various ways, both as a general denominator for a wide array of difficult emotions about climate change and as a more specific term for an emotion in a stricter sense (Kurth & Pihkala, 2022). As an emotion, climate anxiety is evoked by observing threats connected with the climate crisis, and this emotion is fundamental for developing adaptive behaviour (Marks & Hickman, 2023).

In literary criticism, emotions have been observed for decades, but more explicit attention to eco-emotions and climate emotions is a much newer phenomenon. After all, these overarching emotionalities are relatively new, and interdisciplinary research interest in them has grown rapidly (Pihkala, 2020b). As environmental problems, such as global warming, have intensified, so have the numerous manifestations of these emotions become evident in different texts. In the early years of the twenty-first century, the rapidly escalating climate crisis began to have a significant impact on children's and young adult literature and culture. In the past 20 years, writers of fiction for children and young people have studiously sought ways to fight global climate problems and pondered questions of environmental responsibility. As the tenets of environmental education have gained more ground, discussions on issues such as climate change, consumption, and loss of biodiversity have begun to appear in literature written for younger and younger readers.

Several issues in relation to environmental education and psychology have been pondered in connection to fiction for children and young adults. A classic theme is the potential of texts to spark pro-environmental behaviour (PEB). Issues related to emotions, psychological well-being, and functionality are closely connected to PEB. For example, as climate concern and even climate anxiety are increasing among children and young people (e.g. Hickman et al., 2021), many literary scholars have been debating whether the prospects painted by fiction have become too agonizing, and whether the descriptions of the environment have begun to burden young people with too much responsibility and eco-guilt (e.g. Basu et al., 2013; Lahtinen, 2019; Oziewicz & Saguisag, 2021). In other words, literature scholars and teachers who use literature in environmental education have increasingly grappled with how to engage with climate anxiety and other climate emotions constructively in the classroom (e.g. Atkinson & Ray, 2024; Pihkala, 2020a).

Climate Anxiety and the Impacts of Climate Change on Youth

Climate anxiety can be thought of as both an emotion and a wider psychological phenomenon which can include various combinations of emotions (Pihkala, 2020b). For example, there can be feelings of sadness, guilt, and rage combined with climate anxiety (e.g. Jensen, 2019; Pihkala, 2022). Figure 10.1 shows an array of possible climate emotions; this wheel was published in 2023 and is

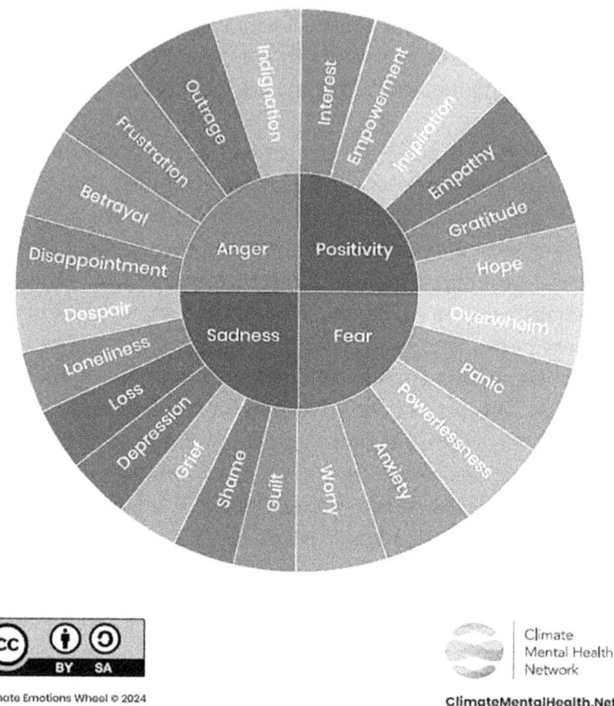

Figure 10.1 Climate Emotions Wheel

used in various settings all over the world, including education (Climate Mental Health Network, 2024).

The subjects of climate emotions can differ. Some of them can be evoked by the climate crisis in its experienced entirety, while some of them are evoked, for example, by collective climate action, such as empowerment and gratitude (Pihkala, 2022). There can be different normative assumptions and "feeling rules" in relation to them in various contexts (e.g. Mosquera & Jylhä, 2022), and it is an important task for education to observe these dynamics (e.g. Bryan, 2024; Oziewicz, 2024). Various fictional and non-fictional texts can be used in education to explore forms of climate anxiety, other climate emotions, and ways in which people react to them (e.g. Pihkala, 2020a).

Youth researchers have been paying attention to climate anxiety among children and young people whose concern about the condition of the planet has been a central topic of discussion and news coverage around the world (Hickman et al., 2021). In particular, the School Strike for Climate movement, started by Greta Thunberg in 2018, incited children and young people to appeal to adults and to demonstrate for finding solutions to the climate crisis (e.g. Bowman, 2019; Ojala, 2023). In the 2020s, with the COVID-19 pandemic and other societal crises in the world, the School Strike for Climate movement experienced difficulties, and there has been a rise in climate disavowal in societies.

Though children and young people experience the climate crisis and react to it in varied ways, studies show that the climate crisis will increasingly affect children's and young adults' health in the future. Various forms of climate anxiety are also predicted to increase (see, e.g. Brophy et al., 2023; Léger-Goodes et al., 2022). In other words, climate issues will be present in more and more complex ways not only in children's and young adult literature but also in the readers' everyday life and well-being. This will affect how texts are used in education: as many scholars have observed, this changes how cli-fi (climate fiction) is understood, since the theme no longer refers to something far in the future (e.g. Bense, 2022; Millet, 2021).

Studies in recent years have shown that the prospects for the future of Finnish young people of 15 to 29 years of age are most clearly overshadowed by climate change, which causes uncertainty and insecurity; this is, of course, an international phenomenon (see, e.g. Whitlock, 2023). As early as 2010, the Finnish Youth Barometer showed that climate change, pollution, and population growth are of even more concern to young people than their personal challenges. The Barometers of 2016 and 2018, in turn, show that respect for the environment, ecological sustainability, and slowing down climate change are ever more important for young people, whereas material values and owning things have lost ground. The environmental ethics of contemporary young people includes, as an intrinsic element, the notion that everyone's actions and even the tiniest choices carry significance. However, at the same time, older generations are still expected to carry out major improvements to achieve a change of course (e.g. Lahtinen, 2019; Piispa & Myllyniemi, 2019, pp. 63–65). The climate question is powerfully a question of generational politics (e.g. Piispa et al., 2021, p. 9), and young people globally have expressed profound disappointment at the climate politics of governments (Hickman et al., 2021).

The action of climate activists is founded on a value-rational ethos, having no clear-cut goals but rather, being guided by a vision of a world that is better, ecologically sustainable, and more equitable. Through their actions, young climate activists are demonstrating that the traditional concepts of education, where the older generation comes across as a mature standard bearer, do not (at least fully) apply in an era of ecological crisis (e.g. Bowman, 2019). By questioning the basic commitments and beliefs of contemporary lifestyles, young people are turning the traditional fostering relationship upside down and setting targets of self-development and growth for older generations (e.g. Lahtinen et al., 2022).

Environmental Education Amidst the Climate Crisis in a Finnish Educational Context

The rapidly growing ecological problems and their interconnections with various other social crises have affected environmental education in profound ways. Environmental education, which includes climate education, refers to all the action that fosters, in individuals and communities, environmental awareness and ways of acting and living that are in line with sustainable development. Environmental education strives to raise awareness of the mutual dependencies of economic, social, political, and ecological factors and to improve both individuals' and communities' procedures related to the environment (Lehtonen et al., 2020; also Värri, 2007, pp. 12–15).

Finland has a long history of environmental education, with roots in the late nineteenth century and various developments during the twentieth century. The climate crisis and environmental responsibility have received emphatic attention both in kindergartens and in schools (Cantell et al., 2020). The teaching in both kinds of institutions is guided by national curricula, which outline the guiding principles and goals of instruction. Based on the national curricula, school-specific curricula are drawn up, but neither document type defines the teaching content in detail. Compulsory school attendance begins at the age of 7 and is divided into elementary school (ages 7–12), lower secondary school (ages 13–15), and upper secondary school (ages 16–18). Before reaching school age, children attend early childhood education and a one-year preschool. Vocational education and voluntary basic education in the arts, provided by art schools, have curricula of their own.

The curricula of all aforementioned stages and types of schools reflect the principles of sustainable development and the associated social, cultural, economic, and ecological dimensions (Finnish National Agency for Education, 2014a, 2014b, 2018, 2019, 2022). These principles are expected to be integrated into each school subject as environmental pedagogical goals. Finnish National Core Curriculum in 2014 was a significant pedagogical opening move towards creating a sustainable relationship between humans and nature. One of the goals of instruction in schools is to enhance *ecosocial education*, which is a holistic educational framework developed by several Finnish scholars and educators (Pulkki et al., 2021). Its starting point is the understanding that humans are fundamental and permanently dependent on each other and on non-human nature, which places a particular responsibility for growth on humanity. Ecosocial education involves understanding the rights and duties of humans within our interdependent relationship with nature and each other, which is achieved through personal growth. Its core values include freedom, responsibility, fairness, and interpersonal connections (Salonen, 2014; Keto et al., 2022). Due to its holistic character, ecosocial education explicitly includes attention to emotions in education, even while the integration of ecosocial education and eco-emotion research is still underway (Pulkki et al., 2021).

Environmental Education in Mother Tongue and Literature Classes

Environmental knowledge and global cultural understanding can also be fostered in literature classes. The Finnish subject "Mother Tongue and Literature" (in Finnish, äidinkieli ja kirjallisuus),[1] in which the students read both fiction and non-fiction texts, offers opportunities to examine issues related to climate emotions, texts, and reading. The teaching of literature, too, is expected to enhance the understanding of environmental challenges and to prepare students to implement sustainable development (see Finnish National Agency for Education, 2019, pp. 9–10, 67; Lahtinen, 2019, p. 75). Both fiction and non-fiction entail at least an implicit notion of the relationship between humans and non-human nature, so that they mould our thoughts on nature (or, the more-than-human world) and environmental issues. It is therefore important to examine different texts in school in relation to their views of the human–nature relationship, the rhetorical means they use, and the eco-emotions they may evoke.

Environmental endeavours in mother tongue and literature also fit nicely with the core objective of the subject, which is the development of literacy, reading, and interpretation skills. One of the current key concepts of Finnish curricula is multiliteracy, which refers to the ability to read, understand, interpret, and write diverse texts in different contexts and with different devices. The concept contains different targeted special reading skills, such as media literacy, health literacy, and economic literacy. The curricula presume that multiliteracy be fostered in all school subjects (Finnish National Agency for Education, 2021). Discussions on environmental education refer to environmental literacy, which can equally be regarded as an aspect of multiliteracy. The concept of environmental literacy (see also Orr, 1992) refers not only to the individual's ability to identify attitudes and actions detrimental to the environment but also to preparedness to act in accordance with the principles of sustainable development (see, e.g. Abiolu & Okere, 2012; McBride et al., 2013).

Eco-emotional Multiliteracy and Reading Skills

Multiliteracy and environmental education have been linked together by various researchers (e.g. Wong & Kumpulainen, 2019), but while emotion-related themes have been discussed, they have not yet been given full attention in research literature. We argue that in the current situation, literacy, reading, and interpretation skills should be developed and used in relation to eco-emotions. We follow the principles of multiliteracy that both reading and writing skills are essentially shared and interactive functions (e.g. Sintonen & Kumpulainen, 2018). Therefore, we apply multiliteracy to include skills of reading, expressing, and interpreting eco-emotions in various kinds of texts and media. The *eco-emotional multiliteracy skills* include the following:

- observing/identifying various eco-emotions in texts;
- recognizing which of eco-emotions the texts evoke in oneself;
- critically analysing which eco-emotions the text may evoke in others, and suggesting possible purposes of the author(s) in relation to eco-emotions; and
- creating and expressing eco-emotionally sensitive, skilful, and powerful texts.

Eco-emotional multiliteracy is intimately bound with ethical and psychological issues. First, students should be supported so that they are able to cope with the eco-emotions evoked or strengthened by the texts. Many educators and researchers have been working on methods related to providing space for peer support and validation, giving coping tips (for example, via psychological guidebooks), and possible trigger warnings. In practice, this requires educating educators and supporting them with their own eco-emotions (e.g. Atkinson & Ray, 2024; Pihkala, 2020a). Second, the various potential outcomes for climate anxiety should be carefully considered. Some psychologists and researchers are optimistic about the possibilities of engaging constructively with climate anxiety via reading how characters in texts experience environmental issues and emotions. For example, psychotherapist Turp (2023) argues that imaginative identification with the characters in various cli-fi situations can help readers think about their own possible responses, the values that guide their behaviour, and their various emotions (pp. 112–113). However, we argue that empirical ecocriticism is needed here, because people can react differently and their skills, including both reading skills and coping skills, affect the outcomes. For some readers, cli-fi may help them engage constructively with climate anxiety and other eco-emotions, while for others dystopias may increase their distress. And for some, no climate emotions whatsoever may be evoked (see Helle et al., 2024; Schneider-Mayerson et al., 2023).

With this agenda of developing eco-emotional multiliteracy skills in mind, we now turn to analysing related themes from the point of view of literature studies, education, and interdisciplinary eco-emotion research. Next, we will discuss tensions related to environmental education and emotions in children's and young adult literature.

Tensions in Environmental Education in Children's and Young Adult Literature

Global environmental concerns are clearly visible in the themes of contemporary children's and young adult literature, but there is a need for critical evaluation of how this literature contributes to environmental and ecosocial education. Though the tendencies for environmental education through literature are not new, the scale of the environmental problems dealt with in books directed to young readers has grown significantly (see, e.g. Oziewicz & Saguisag, 2021; Lahtinen, 2013, pp. 102–104). During the last decade, a considerable amount of fiction providing support for children's and young adults' agency has been published (e.g. Rönkä, 2020, p. 6) and used for environmental education in kindergartens and schools, but there is a need for research regarding the impacts of various ways in which it is used.

Geraldine Massey and Clare Bradford (2011) reflect on the environmental pedagogical potential of children's and young adult literature and note that environmental literature not only reflects local or global changes but also strives to socialize children and young people into a way of life that respects nature, which is a classic tenet in environmental education. The values offered to children and young people by literature are based on the world views and values of the adults who write the books. If the young readers even partly accept the reader role offered to them in these books, they may adopt the world view conveyed by them. In an environmental framing, the reading of fiction will enhance their understanding of a way of life that takes environmental values into account (Ibid.). In this way, literature may help to encourage children and young people to act for the good of the environment, to ponder the problems of the planet, and to imagine different futures that the current development and endangering of nature may lead to in time (e.g. Massey & Bradford, 2011; Wong & Kumpulainen, 2019). Just like other forms of environmental and climate education, children's and young adult literature may foster environmental agency and ecosocial education in its readers.

However, there are several tensions inherent in this. First, learning about environmental damage can increase not only constructive worry and anxiety but also paralysing forms of them (Pihkala, 2020a, 2024). Second, from the viewpoint of ecosocial education, children's and young adult literature contains an inherent tension in relation to socialization. On the one hand, it has been considered as promoting a middle-class social order, while on the other hand, it is seen as containing revolutionary elements that incite readers to challenge the conventional norms of the adult world (see Laakso et al., 2011, pp. 10, 15–18; Lurie, 1990, pp. x–xi; Mickenberg, 2006, p. 7). Similar tensions between individuation and societalization have been addressed in discussions of philosophies of education. Education aims at learning and societalization, with which individuals acquire the knowledge, skills, habits, code of conduct, values, and attitudes that enable them to function as members of society; that individuals grow into independent agents capable of creative action (e.g. Siljander, 2014, pp. 46–53; Värri, 2007, p. 70).

The dilemma of socializing the student into a system that is fundamentally being called into question pertains to climate education as well: how can a socialization that is supporting the climate crisis be promoted in a society where environmental problems are an integral part of the technologization and the capitalistic economic system which drives the destructiveness (see Värri, 2018, pp. 24–25)? This dilemma can cause climate anxiety for educators who recognize it. Many young people struggle with the magnitude of changes needed for contemporary societies to achieve ecological sustainability (Atkinson & Ray, 2024). On the positive side, it is possible to engage students in this dilemma by using diverse texts of fiction and non-fiction as starting points for discussions, and thus educating eco-emotional literacy skills at the same time. For example, the following fundamental question can be discussed with students: What kind of socialization does the text proffer, and is this aligned with the societal changes which are needed? In relation to eco-emotions, this resonates with the issue of whether the text educates into a position of eco-guilt and/or a position of eco-anger, of moral outrage against unjust systems.

Children's and Young Adult Literature in the Framework of Eco-emotions

These controversies between communality and individualization, and between responsibility and freedom are reflected in the pastoral tradition of children's and young adult literature, in which child characters have adventures in nature, outside home. In Western children's literature, natural milieux often offer children freedom from the world adults govern unjustly; the world whose technologization, laws, and cruelty are typically criticized in children's literature. The child character itself may also be a pastoral figure, in which case the flight from the world ruined by adults functions as an allegory for regeneration and new prospects (e.g. Natov, 2003, pp. 91–92). This tradition, in turn, has been influenced by the philosophical thinking of Jean-Jacques Rousseau, in which the child represents pristine nature. Thus, in children's and young adult literature, nature often serves as a metaphor for the inner goodness of humans and the idealized freedom of childhood and as a milieu in which the child's natural instincts and society's rules can be reconciled (see Byrnes, 1995, pp. 12–13).

Pastorals, which are essentially based on juxtaposition of the city and the countryside, have become highly controversial amidst the climate crisis. Global warming affects all kinds of terrains, and youth who experience ecological grief and/or climate anxiety sometimes have difficulties in finding solace in nature, because natural environments may remind them of all the threats and destruction (Hickman, 2020). However, finding relief in nature is still a theme in contemporary literature, and the child and the young person are still often idealized. According to Rönkä (2020), the environmental problems described in contemporary literature are shown to derive from the greed and selfishness of the adult characters, whereas the agency and resourcefulness of children, fairy-tale figures, and animals are accentuated.

Overall, children's books that teach environmental values and agency have grown into a literary phenomenon in Western countries. Finnish children's literature, too, has shown diligent effort to teach little ones practical skills to protect biodiversity and slow down the climate change. Recycling, for example, has grown into a pedagogical theme of its own in picture books that teach children to make consumption-critical choices as a responsibility and an obligation, as well as a positive mode of action (e.g. Hytönen, 2021; Lahtinen, 2013). Such picture books show graphically how children, on their own, can work for the good of the environment in their concrete everyday action. Such books have been criticized for lack of societal perspective as they do not often promote criticism of consumerism or deliberations on the sorts of economic, cultural, or social problems that attach to waste (Christenson, 2009; Muthukrishnan & Kelley, 2017). Environmental education scholars have noted that in many schools, the transition from individualistic environmental actions to civil action against systemic problems is a very difficult one (Cantell et al., 2020).

In young adult literature, a new subtype has been formed by ecological dystopias, which deal with topics such as drought, storms, availability of drinking water, and climate exile (Laakso et al., 2019; Rönkä, 2020). Placing the events in the near future provides scope for an adventurous plot, but the books also offer clear lessons and warnings about the possible consequences of current choices. Pedagogically, the dystopias invite the reader to ponder on similarities between the current and the fictitious future. They often couple the maturation story of the young protagonist(s) with social awakening (Hinz & Ostry, 2003, p. 9). However, as Millet (2021) points out, the distinctions between cli-fi and present conditions become increasingly hazy, since so many areas in the world already suffer from climate disruption, and this should be considered when using cli-fi in the classroom.

Dystopias also remind us how children and young adults may repel the environmental pedagogical reader positions that adults offer them. It has been observed that dystopias for young adults may make politically motivated violence more acceptable among the readers (Jones & Paris, 2018). Similarly, the pedagogics of the strongly individualistic or heroic story template have been criticized for excessively emphasizing the inner change in the individual instead of the social turning point, even though the solving of global environmental problems requires communal solutions and stories (e.g. Moriarty, 2021).

Cli-fi and dystopias which emphasize heroic youth can also promote the problematic discourses related to socialization and responsibility: one may ask whether it is ethically justifiable for books written by adults to shift the pressures of solving the climate crisis onto future generations, furthering their climate anxiety and guilt. Young people themselves often criticize the ways in which adults talk about how the youth will save the world (e.g. Diffey et al., 2022). To recognize the ethical problems and controversies entailed in environmentally oriented fiction young readers need eco-emotional multiliteracy skills. Then again, pondering the problems calls for narrative agency, which refers to the ability to use, interpret or reinterpret different cultural narratives and make use of them in critical assessments of our own life and the world around us (e.g. Meretoja, 2018). The strengthening of narrative agency is based on multiliteracy, for the more diverse the literacy, the more critically can children and young adults relate to their reading and question the role models or reader positions offered to them.

Reading Literature as a Part of Environmental Education

From the viewpoint of the environmental-pedagogical goals in mother tongue and literature and in early childhood education, the biggest problem is not a lack of environment- and climate-related fiction. It is the eroding reading skills and the diminishing reading motivation of children and young adults. The reading of literature may stimulate the readers to action, empower them, and help them to deal with their environmental anxiety, but children and young adults now read less and less. This is a widespread international phenomenon (e.g. Clark et al., 2023). In Finland, it is the 10- to 14-year-olds who read the most books in their free time, whereas the 15- to 24-year-olds read the fewest (Hanifi, 2021). In the Programme for International Student Assessment (PISA) 2018 study, as many as 63% of Finnish ninth-grade boys reported that they would read only if they had to (Leino et al., 2019). Among students in the elementary school, reading has also decreased; for example, according to a comparative study conducted in 2016, one-third of Finnish fourth-graders did not read fiction in their free time (Leino et al., 2017). When free time reading decreases, the literature education in schools grows in importance: for many a young adult, the books they read at school become the only gateway to the world of literature.

The waning of reading habits and the weakening of young adults' basic reading skills are connected. In the PISA 2018 study, about 14 percent of Finnish 15-year-olds were weak readers (Leino et al., 2019), and in the PISA 2022 study, reading skills of weak readers continued to deteriorate. This is an international trend (OECD, 2023). Being a weak reader means that the young adult's reading and writing skills are inadequate for completing their studies without problems. In addition, weak readers will struggle to understand and compose texts used and required in society. Frail literacy skills weaken the person's study motivation and are connected to the threat of unemployment and social exclusion (Arffman & Nissinen, 2015; Leino et al., 2019; Vettenranta et al., 2016). In essence, literacy is an issue of social equality and a mainstay of the democratic society. Young people with weak literacy skills have fewer opportunities for civic participation and, therefore, are less equipped to deal with and take action on environmental issues. The ecosocial education and responsibility pursued in the curricula presume skills of critical reading, civic engagement, and skills of interaction in this world of global climate problems, but these aims cannot be reached when the reading skills and motivation are low (Albrecht et al., 2021; Grünthal, 2020; Leino et al., 2019).

Kelly Hübben (2017) emphasizes the role of critical reading in environmental education. The reading of fiction requires special reading skills that enable the reader to understand multi-layered and elliptical texts. The teaching and training of these skills is among the core missions of literature education at school (Grünthal, 2020). In connection with fiction, one also talks about close reading, which traditionally refers to precise reading that focuses on the structures, poetics, and language of the text. Recently, practitioners of close reading have been reminding us that reading always takes place in a certain sociocultural and political context. Eco-criticism and post-humanism, for example, have made use of close reading to understand the linkage humans have with the environment and have thereby charted the boundaries of human understanding (e.g. Kortekallio & Ovaska, 2020). By developing critical close reading strategies, students can contemplate the complex power dynamics that are typical for children's and young adult literature. These dynamics often involve binaries, such as child/adult, nature/culture, or human/non-human relationships.

However, reading fiction is increasingly more difficult for many young readers, for it requires absorption. Long-span reading is characterized by deliberation, empathizing, and immersing oneself in the text, to the extent that even the sense of time and place may vanish; this sort of reading is also referred to as slow and immersive reading (Knuuttila et al., 2021; Kortekallio & Ovaska, 2020; Wolf & Barzillai, 2009). Studies indicate that it is the reading of long texts and fiction that best develops reading skills and that the amount of fiction young adults read correlates with their learning achievements in other school subjects (Harjunen & Rautopuro, 2015). For many a young adult, reading a book is perceived as a laborious and old-fashioned activity. In spring 2019, teachers of mother tongue and literature in Finnish lower secondary schools reported that the latest assignment of reading a whole book caused difficulty for as many as 80% of the students (Aaltonen, 2019; Heikkilä-Halttunen, 2015, pp. 213–219).

Indeed, it is among the biggest challenges in using children's and young adult literature as an instrument of environmental education that instead of fiction, many of today's young adults read mostly short online texts such as social media, headlines, and advertisements, which are typically just skimmed (e.g. Hanifi, 2021; Saarenmaa, 2021). We argue that skimming and reading mostly short texts deteriorate students' capability to manage climate change and anxiety, which call for understanding of complex phenomena and diverse interrelations. These, in turn, are typical features in novels and long texts.

Methods for Developing Reading Skills and Eco-emotional Multiliteracy

One solution for developing both students' reading skills and environmental pedagogy addressing climate anxiety could be methods of shared reading, which are founded on interactivity and enhance the motivation to read. We assume that shared reading may also be more favourable in dealing with climate anxiety than solitary reading, because it includes sharing emotions, thoughts, and visions with other students. However, studies show that the methods favoured in Finnish elementary and lower secondary school literature education are still mostly individual ones, such as oral presentations, book reports, and essays (on elementary school, see Grünthal et al., 2019, pp. 168–169; on lower secondary school, see Aaltonen, 2020).

In her doctoral dissertation dealing with drama work as a tool for climate education, Anna Lehtonen (2021) demonstrates how drama can improve the perception of factual connections, continua, and different points of view, which is precisely the ability required for understanding climate issues. By means of drama, one can envision different futures and achieve a kind of listening, contact-strengthening state, which may induce critical thinking and a will to act responsibly. This can also help in encountering climate anxiety (Lehtonen & Pihkala, 2021). Such a state is not unlike the state discussed in connection with the eco-critical and post-humanistic close reading outlined by Kortekallio and Ovaska (2020, pp. 62–65). By this sort of close reading, they mean a reading that not only analyses the textual structures and the language but also adopts a listening and interacting approach and takes the physical, cultural, emotional, and social dimensions of reading into consideration. If this sort of shared and verbalized close reading creates contacts with other readers, the author, and the non-human environment, it may also serve as an instrument of social change. As stated earlier, we call this ability eco-emotional multiliteracy.

Similarly, reading circles, referred to by a variety of names in international research, have been found by teachers to be a good method for inspiring students to read, for they create a commitment and motivation in young people who are unaccustomed to reading and enable shared deliberation of themes that are of significance in their lives. They also build up involvement and language skills (Aaltonen, 2020; Granqvist, 2023). In schools, reading circles and shared reading are teacher-led projects in which small groups meet to discuss a book they have all read and to do exercises assigned based on the book. For this reason, they also enable teachers to monitor and address climate anxiety in their students. The discussions diversify the reading experiences of the participants, enhance their interpretation skills, and guide them towards considering the viewpoints of others (Langer, 2010; Olin-Scheller & Tengberg, 2012). Different forms of communality are important for young adults, for both blending in with their peers and developing their own identity. Thus, just like drama, the reading circles offer a method for developing the students' emotional skills and imagination for the purposes of climate education and eco-emotional multiliteracy (Bowman et al., 2024).

The heaviest responsibility for acquainting the students with fiction and teaching them literacy skills is borne by mother tongue and literature teachers, although multiliteracy skills should be taught in all school subjects. Regardless of the method used, the narrative agency and environmental, eco-emotional multiliteracy is also fostered by research-based factual information about the environment and sustainable development. Eco-emotional multiliteracy continuously relates the fictive world of the text to the current state of the environment and the current debate about it on the one hand and to such potential futures, on the other hand, as fiction can create, and the reader can envision further.

Conclusions: Towards Eco-emotional Multiliteracy

Regardless of the era, school pedagogy and the curricular guidelines are based on the learning contents and teaching goals defined as the most important in the debates and struggles about school curricula carried on by the different interest groups of the society. These learning contents and teaching goals always contain both conscious and subconscious educational philosophical views about the foremost value goals of school pedagogy. On the road from the country school curriculum (MOPS, 1925) to modernity, the main focus has shifted from local to global education, from Christian civic education to ecosocial education (Saari et al., 2017).

According to Outi Hytönen (2021), good children's and young adult literature always includes several levels instead of mere didacticism. Indeed, several literary scholars have expressed concern about the sort of attitude training that emphasizes children's and young adults' responsibility and the sort of utilitarian thinking that narrows the roles and purports of literature (e.g. Seutu, 2020). Opposing voices argue that global environmental problems attach so closely and in so many ways to issues of environmental justice that children's and young adult literature should ever more resolutely address consumership and the climate crisis, with which the new generations will be wrestling throughout their lives (see Oziewicz & Saguisag, 2021, pp. vii–ix). From the educational point of view manifested in the curricula, the issue is seen differently: in the context of school teaching, every literary text obtains pedagogical significance regardless of the book's artistic value.

The pedagogical goal of school teaching is the attainment of multileveled and diverse multiliteracy, which also develops the reader's imaginativeness. Without that, there is no hope and no vistas for alternative futures. This goal is well served by communal methods of literature education, such as reading circles and drama, which entail sharing of reading experiences and discussing them. While Finnish schools still favour individual literature assignments, it is precisely the communal methods that succeed in generating more reading motivation in weak readers, besides which they provide support for dealing with emotional experiences and tackling climate anxiety. Communal strategies in literature education do not seek to find a consensus about the correct way of interpreting literature but rather to deepen the participants' reading experience and their ability to understand different points of view, and to provide the sort of ecosocial education referred to in the goals of the curricula.

One attempt to solve the contradiction between individuation and societalization entailed in environmental education is the recent climate movement, in which the young act in close adherence to the way of life of their own society but strive at the same time to question and dismantle those of its beliefs and commitments that are ecologically unsustainable (see Lahtinen et al., 2022, pp. 51–54; Siljander, 2014, pp. 49, 50). At best, the critical reading of literature offers arenas for questioning, alternative worlds, and diverse perspectives. Reading climate fiction and discussing it may carry significance in children's and young adults' lives, for it can change both the individual's and the community's attitudes to the environmental crisis. Fostering students' ecosocial education and strengthening their eco-emotional multiliteracy offers them capabilities for dealing with environmental responsibility, various climate emotions, including anxiety, and of taking action – for example, writing and sharing environment-oriented texts.

Communication between the philosophy of education, study of literature, empirical reader research, and pedagogy needs to be strengthened for the benefit of the ecosocial aspirations of the curricula and the environmental pedagogical emphases of children's and young adult literature. The creation of literature and readers' interaction with it are affected by many cultural and narrative models, including the ideals and aspirations generated in environmental philosophy and the philosophy of education. Furthermore, books that are environmentally aware and promote ecosocial education may also contain thought patterns that are contradictory and may easily go unnoticed in routine reading. For identifying such thought patterns, critical reading skills are required, and educational discussions can help. In the theoretical presumptions of the study of literature and environmental pedagogy, too little attention has been paid to the question of how sufficient the reading skills of the average child or young adult are for interpreting and receiving the message of the books written by adults.

We argue that increasing the amount of literature read in schools and promoting eco-emotional multiliteracy skills enhance students' imagination and creativity, which, in turn, gives them capacities to imagine alternative futures, vision new solutions and tackle wicked problems. Together with methods of shared reading, book talks, and drama, they give students tools to cope with their emotions, fears, and climate anxiety, and they open up the window of hope.

Note

1 Because the concept of mother tongue is used in Finnish curricula, we follow their example. By mother tongue and literature education, we mean L1 (first language) education, in our case education in Finnish language and literature. In Finnish curricula, however, mother tongue can also be defined as Swedish, Sami, Roma language, sign language, or some other native language spoken by the students.

References

Aaltonen, L. (2019). "Miks mä lukisin kirjaa?" Digitaalinen lukukokemus nuoren lukumotivaation herättäjänä. *Nuorisotutkimus, 3–4*, 69–83. http://hdl.handle.net/10138/324481

Aaltonen, L. (2020). Lukuklaani-tutkimus. Ala- ja yläkoulujen alkukartoitusten koosteet, analyysi ja tiedottaminen. *Tutkimusraportti 2017–2019*. Suomen kulttuurirahasto http://hdl.handle.net/10138/327030

Abiolu, O. A., & Okere, O. O. (2012). Environmental literacy and the emerging roles of information professionals in developing economies. *IFLA Journal, 38*(1), 53–59. https://doi. org/10.1177/0340035211435070.

Albrecht, E., Leppäkoski, E., Vaara, E., Meriläinen, N., & Viljanen, J. (2021). Nuorten osallistuminen ilmastolain valmisteluun. *Nuorisotutkimus, 39*(2), 46–62.

Arffman, I., & Nissinen, K. (2015). Lukutaidon kehitys PISA-tutkimuksissa. In J. Välijärvi & P. Kupari (Eds.), *Millä eväillä osaaminen uuteen nousuun? PISA 2012 – tutkimustuloksia* (Vol. 6, pp. 28–49). Opetus- ja kulttuuriministeriön julkaisuja.

Atkinson, J., & Ray, S. J. (Eds.). (2024). *The existential toolkit for climate justice educators: How to teach in a burning world*. University of California Press.

Basu, B., Broad, K. R., & Hintz, C. (2013). Introduction. In B. Basu, K. R. Broad & C. Hintz (Eds.), *Contemporary dystopian fiction for young adults. Brave new teenagers* (pp. 1–15). Routledge.

Bense, K. (2022). As the climate changes, climate fiction is changing with it. *Inside Climate News*. https://insideclimatenews.org/news/17122022/warming-trends-climate-fiction/

Bladow, K. A., & Ladino, J. (Eds.). (2018). *Affective ecocriticism: Emotion, embodiment, environment*. University of Nebraska Press.

Bowman, B. (2019). Imagining future worlds alongside young climate activists: A new framework for research. *Fennia: International Journal of Geography, 197*(2), 295–305.

Bowman, B., Germaine, C., Kishinani, P., & Balchin, C. (2024). The climate imaginary: Reading fiction to make sense of the climate crisis. In J. Atkinson & S. J. Ray (Eds.), *The existential toolkit for climate justice educators: How to teach in a burning world* (pp. 285–292). University of California Press.

Brophy, H., Olson, J., & Paul, P. (2023). Eco-anxiety in youth: An integrative literature review. *International Journal of Mental Health Nursing, 32*(3), 633–661. https://doi. org/10.1111/inm.13099

Bryan, A. (2024). The social ecology of responsibility: Navigating the epistemic and affective dimensions of the climate crisis. In J. Atkinson & S. J. Ray (Eds.), *The existential toolkit for climate justice educators: How to teach in a burning world* (1st ed., pp. 201–209). University of California Press.

Byrnes, A. (1995). *The child. An archetypal symbol in literature for children and adults*. Peter Lang.

Cantell, H., Aarnio-Linnanvuori, E., & Tani, S. (2020). *Ympäristökasvatus. Kestävän tulevaisuuden käsikirja*. Santalahti.

Christenson, M. A. (2009). Children's literature on recycling: What does it contribute to environmental literacy? *Applied Environmental Education and Communication, 7*(4), 144–154. https://doi.org/10.1080/15330150902744160.

Clark, C., Picton, I., & Galway, M. (2023). Children and young people's reading in 2023. *National Literacy Trust*. https://literacytrust.org.uk/

Climate Mental Health Network. (2024). *Climate emotions wheel*. www.climatementalhealth.net/wheel

Diffey, J., Wright, S., Uchendu, J. O., Masithi, S., Olude, A., Juma, D. O., Anya, L. H., Salami, T., Mogathala, P. R., Agarwal, H., Roh, H., Aboy, K. V., Cote, J., Saini, A., Mitchell, K., Kleczka, J., Lobner, M. G., Ialamov, N., Borbely, M., . . . Lawrance, E.

(2022). "Not about us without us" – the feelings and hopes of climate-concerned young people around the world. *International Review of Psychiatry*, 34(5), 499–509. https://doi.org/10.1080/09540261.2022.2126297

Finnish National Agency for Education. (2014a). *Esiopetuksen opetussuunnitelman perusteet*. www.oph.fi/sites/default/files/documents/esiopetuksen_opetussuunnitelman_perusteet_2014.pdf

Finnish National Agency for Education. (2014b). *Perusopetuksen opetussuunnitelman perusteet*. www.oph.fi/sites/default/files/documents/perusopetuksen_opetussuunnitelman_perusteet_2014.pdf

Finnish National Agency for Education. (2018). *Varhaiskasvatussuunnitelman perusteet*. www.oph.fi/sites/default/files/documents/varhaiskasvatussuunnitelman_perusteet.pdf

Finnish National Agency for Education. (2019). *Lukion opetussuunnitelman perusteet*. www.oph.fi/sites/default/files/documents/lukion_opetussuunnitelman_perusteet_2019.pdf

Finnish National Agency for Education. (2021). *National literacy strategy 2030*. Publications, Finnish National Agency for Education. Retrieved October 18, 2024, from https://www.oph.fi/sites/default/files/documents/National_literacy_strategy_2030.pdf

Finnish National Agency for Education. (2022). *Kestävä kehitys kasvatuksen ja koulutuksen normiperustassa*. Retrieved June 12, 2022, from www.oph.fi/fi/opettajat-ja-kasvattajat/kestava-kehitys-kasvatuksen-ja-koulutuksen-normiperustassa

Granqvist, A. (2023). How I live [participate] now: Re-negotiating social belonging and linguistic participation in book-group discussions. *European Journal of Applied Linguistics*, 11(2), 460–487. https://doi.org/10.1515/eujal-2023-0003

Grünthal, S. (2020). Lukutaito ja lukeminen. In L. Tainio, M. Ahlholm, S. Grünthal, S. Happonen, R. Juvonen, U. Karvonen & S. Routarinne (Eds.), *Suomen kieli ja kirjallisuus koulussa* (pp. 165–206). Suomen ainedidaktinen tutkimusseura. http://hdl.handle.net/10138/316123 62

Grünthal, S., Hiidenmaa, P., Routarinne, S., Satokangas, H., & Tainio, L. (2019). Alakoulun kirjallisuuskasvatusta kartoittamassa: Lukuklaanin opettajakyselyn tuloksia. In M. Rautianen & M. Tarnanen (Eds.), *Tutkimuksesta luokkahuoneisiin* (pp. 161–180). Suomen ainedidaktinen tutkimusseurary. http://hdl.handle.net/10138/298542

Hanifi, R. (2021). Lukeminen muutoksessa. In R. Hanifi, J. Haaramo & K. Saarenmaa (Eds.), *Mitä kuuluu vapaa-aikaan? Tutkimus, tieto ja tulkinnat* (pp. 133–157). Tilastokeskus. https://urn.fi/URN:ISBN:978-952-244-683-1

Harjunen, E., & Rautopuro, J. (2015). *Kielenkäytön ajattelua ja ajattelun kielentämistä. Äidinkielen ja kirjallisuuden oppimistulokset perusopetuksen päättövaiheessa 2014: keskiössä kielentuntemus ja kirjoittaminen*. Kansallinen koulutuksen arviointikeskus. https://omalinja.fi/wp-content/uploads/2016/03/KARVI_08151.pdf

Heikkilä-Halttunen, P. (2015). *Lue lapselle! Opas lasten kirjallisuuskasvatukseen*. Atena.

Helle, A., Lahtinen, T., Löytty, O., & Pihkala, P. (2024). Lukiolaiset Teemestarin kirjaa lukemassa: Ilmastokirjallisuuden herättämiä ajatuksia ja tunteita. *Kulttuurintutkimus*, 41.

Hickman, C. (2020). We need to (find a way to) talk about . . . Eco-anxiety. *Journal of Social Work Practice*, 34(4), 411–424. https://doi.org/10.1080/02650533.2020.1844166

Hickman, C., Marks, E., Pihkala, P., Clayton, S., Lewandowski, R. E., Mayall, E. E., Wray, B., Mellor, C., & Susteren, L. van. (2021). Climate anxiety in children and young people and their beliefs about government responses to climate change: A global survey. *The Lancet Planetary Health*, 5(12), e863–e873. https://doi.org/10.1016/S2542-5196(21)00278-3

Hinz, C., & Ostry, E. (2003). Introduction. In C. Hinz & E. Ostry (Eds.), *Utopian and dystopian writing for children and young adults* (pp. 1–22). Routledge.

Hübben, K. (2017). *A genre of animal hanky-panky? Animal representations, anthropomorphism and interspecies relations in the little golden books.* Stockholm University.
Hytönen, O. (2021). Kuvakirjat nyt: Ympäristöasiaa ja tunteista taidetta. *Parnasso, 71*(1), 30–37. https://jyu.finna.fi/Record/arto.017521313/HREF
Jensen, T. (2019). *Ecologies of guilt in environmental rhetorics.* Palgrave Macmillan.
Jones, C. W., & Paris, C. (2018). It's the end of the world and they know it: How dystopian fiction shapes political attitudes. *Perspectives on Politics, 16*(4), 969–989. https://doi.org/10.1017/S1537592718002153.
Keto, S., Foster, R., Pulkki, J., Salonen, A. O., & Värri, V.-M. (2022). Ekososiaalinen kasvatus: Viisi teesiä ratkaisuehdotuksena antroposeenin ajan haasteeseen. *Kasvatus & Aika, 16*(3), 49–69. https://doi.org/10.33350/ka.111741
Knuuttila, S., Niemi-Pynttäri, R., & Tulonen, U. (2021). Tämä lukemisesta tiedetään: Lukuhaluttomuuden syyt. *Lukuluxsivusto.* Retrieved January 23, 2022, from www.lukulux.fi/tama-lukemisesta-tiedetaan/lukuhaluttomuuden-syyt/
Kortekallio, K., & Ovaska, A. (2020). Lähiludkeminen ennen ja nyt: Ruumiillisia, ympäristöllisiä ja poliittisia näkökulmia. *AVAIN – Kirjallisuudentutkimuksen Aikakauslehti, 17*(3), 52–69. https://doi. org/10.30665/av.95530.
Kurth, C., & Pihkala, P. (2022). Eco-anxiety: What it is and why it matters. *Frontiers in Psychology, 13.* https://doi.org/10.3389/fpsyg.2022.981814
Laakso, M., Lahtinen, T., & Heikkilä-Halttunen, P. (2011). Johdatus lasten- ja nuortenkirjallisuuden luontoon. In M. Laakso, T. Lahtinen & P. Heikkilä-Halttunen (Eds.), *Tapion tarhoista turkistarhoille. Luonto suomalaisessa lasten- ja nuortenkirjallisuudessa* (pp. 9–28). SKS.
Laakso, M., Lahtinen, T., & Samola, H. (2019). Young saviors and agents of change power, environment, and gender in contemporary Finnish young adult dystopias. *Utopian Studies, 30*(2), 193–213. http://doi.org/10.5325/utopianstudies.30.2.0193
Lahtinen, T. (2013). Ilmastonmuutos lintukodossa. In M. Hallila (Ed.), *Suomen nykykirjallisuus 2. Kirjallinen elämä ja yhteiskunta* (pp. 95–106). SKS.
Lahtinen, T. (2019). Ilmastonmuutos kirjallisuudentutkimuksessa ja -opetuksessa. In M. Murto (Ed.), *Kiinni fiktioon. Kirjallisuuden tutkimuksesta ja opetuksesta* (pp. 75–82). Äidinkielen opettajain liitto.
Lahtinen, T., Grünthal, S., & Värri, V.-M. (2022). Kirjallisuus ja lukeminen ympäristökasvatuksessa. In T. Lahtinen & O. Löytty (Eds.), *Joutsen/Svanen. Kotimaisen kirjallisuudentutkimuksen vuosikirja 2022: Empiirinen ekokritiikki* (pp. 51–66). Helsingin yliopisto.
Langer, J. (2010). *Envisioning literature. Literary understanding and literature instruction* (2nd ed.). Teachers College, Columbia University.
Léger-Goodes, T., Malboeuf-Hurtubise, C., Mastine, T., Généreux, M., Paradis, P.-O., & Camden, C. (2022). Eco-anxiety in children: A scoping review of the mental health impacts of the awareness of climate change. *Frontiers in Psychology, 13.* www.frontiersin.org/articles/10.3389/fpsyg.2022.872544
Lehtonen, A. (2021). *Drama as an interconnecting approach for climate change education* (118) [Doctoral dissertation, University of Helsinki, Helsingin yliopisto: Kasvatustieteellinen tiedekunta]. http://urn.fi/URN:ISBN:978-951-51-7445-1
Lehtonen, A., & Pihkala, P. (2021). Encounters with climate change and its psychosocial aspects through performance making among young people. *Environmental Education Research, 27*(5), 743–761. https://doi.org/10.1080/13504622.2021.1923663
Lehtonen, A., Salonen, A. O., & Cantell, H. (2020). Climate change education: A new approach for a world of wicked problems. In J. W. Cook (Ed.), *Sustainability, human well-being, and the future of education* (pp. 339–374). Palgrave Macmillan. https://doi.org/10.1007/978-3-319-78580-6_11

Leino, K., Ahonen, A. K., Hienonen, N., Hiltunen, J., Lintuvuori, M., Lähteinen, S., Lämsä, J., Nissinen, K., Nissinen, V., Puhakka, E., Pulkkinen, J., Rautopuro, J., Sirén, M., Vainikainen, M.-P., & Vettenranta, J. (2019). *PISA 18: ensituloksia. Suomi parhaiden joukossa.* Opetus- ja kulttuuriministeriö. Opetus- ja kulttuuriministeriön julkaisuja, 40. http://urn.fi/URN:ISBN:978-952-263-678-2

Leino, K., Nissinen, K., Puhakka, E., & Rautopuro, J. (2017). Lukutaito luodaan yhdessä. Kansainvälinen lasten lukutaitotutkimus (PIRLS 2016). *Koulutuksen tutkimuslaitos.* http://urn.fi/URN:ISBN:978-951-39-7292-9

Lurie, A. (1990). *Don't tell grownups. The subversive power of children's literature.* Little, Brown and Company.

Marks, E., & Hickman, C. (2023). Eco-distress is not a pathology, but it still hurts. *Nature Mental Health, 1*(6), Article 6. https://doi.org/10.1038/s44220-023-00075-3

Massey, G., & Bradford, C. (2011). Children as ecocitizens: Ecocriticism and environmental texts. In K. Mallan & C. Bradford (Eds.), *Contemporary children's literature and film: Engaging with theory* (pp. 109–126). Palgrave Macmillan. https://doi.org/10.1007/978-0-230-34530-0_7

McBride, B. B., Brewer, C. A., Berkowitz, A. R., & Borrie, W. T. (2013). Environmental literacy, ecological literacy, ecoliteracy: What do we mean and how did we get here? *Ecosphere, 4*(5), 1–20. https://doi.org/10.1890/ES13-00075.1

Meretoja, H. (2018). *The ethics of storytelling. Narrative hermeneutics, history, and the possible.* Oxford University Press.

Mickenberg, J. L. (2006). *Learning from the left. Children's literature, the cold war, and radical politics in the United States.* Oxford University Press.

Millet, L. (2021). Climate crisis is here; so is climate fiction. Don't you dare call it a genre. *Los Angeles Times.* www.latimes.com/entertainment-arts/books/story/2021-07-07/climate-crisis-is-here-so-is-climate-fiction-dont-you-dare-call-it-a-genre

MOPS. (1925). *Maalaiskansakoulun opetussuunnitelma. Komiteanmietintö.* Valtioneuvoston kirjapaino.

Moriarty, S. (2021). Modeling environmental heroes in literature for children: Stories of youth climate activist greta thunberg. *The Lion and the Unicorn, 45*(2), 192–210. https://doi.org/10.1353/uni.2021.0015

Moser, S. (2015). Whither the heart(-to-heart)? Prospects for a humanistic turn in environmental communication as the world changes darkly. In A. Hansen & R. Cox (Eds.), *Handbook on environment and communication* (pp. 402–415). Routledge.

Mosquera, J., & Jylhä, K. M. (2022). How to feel about climate change? An analysis of the normativity of climate emotions. *International Journal of Philosophical Studies, 30*(3), 357–380. https://doi.org/10.1080/09672559.2022.2125150

Muthukrishnan, R., & Kelley, J. E. (2017). Depictions of sustainability in children's books. *Environment, Development and Sustainability, 19*(3), 955–970.

Natov, R. (2003). *The Poetics of Childhood.* Routledge.

OECD. (2023). *PISA 2022 results (volume I): The state of learning and equity in education.* PISA, OECD Publishing. https://doi.org/10.1787/53f23881-en

Ojala, M. (2023). How do children, adolescents, and young adults relate to climate change? Implications for developmental psychology. *European Journal of Developmental Psychology, 20*(6), 929–943. https://doi.org/10.1080/17405629.2022.2108396

Olin-Scheller, C., & Tengberg, M. (2012). "If it ain't true, then it's just a book!" The reading and teaching of faction literature. *Journal of Research in Reading, 35*(2), 153–168. https://doi.org/10.1111/j.1467-9817.2010.01453.x.

Orr, D. W. (1992). *Ecological literacy: Education and the transition to a postmodern world.* State University of New York Press.

Oziewicz, M. (2024). Beyound the accountability paradox: Climate guilt and the systemic drivers of climate change. In J. Atkinson & S. J. Ray (Eds.), *The existential toolkit for climate justice educators: How to teach in a burning world* (pp. 210–217). University of California Press.

Oziewicz, M., & Saguisag, L. (2021). Introduction: Children's literature and climate change. *The Lion and the Unicorn, 45*(2), vx–iv. https://doi.org/10.1353/uni.2021.0011

Pihkala, P. (2020a). Eco-anxiety and environmental education. *Sustainability, 12*(23), 10149. https://doi.org/10.3390/su122310149

Pihkala, P. (2020b). Anxiety and the ecological crisis: An analysis of eco-anxiety and climate anxiety. *Sustainability, 12*(19), 7836. https://doi.org/10.3390/su12197836

Pihkala, P. (2022). Toward a taxonomy of climate emotions. *Frontiers in Climate, 3.* https://doi.org/10.3389/fclim.2021.738154

Pihkala, P. (2024). Emotional and affective issues in environmental and sustainability education. *Oxford Bibliographies Online in Education.* https://doi.org/10.1093/obo/9780199756810-0310

Piispa, M., Kiilakoski, T., & Ojajärvi, A. (2021). Mikä ilmastoliike? Nuorten ilmastoaktivistien poliittinen toimijuus ja ilmastokeskustelut. *Nuorisotutkimus, 39,* 8–26.

Piispa, M., & Myllyniemi, S. (2019). Nuoret ja ilmastonmuutos. Tiedot, huoli ja toiminta Nuorisobarometrien valossa. *Yhteiskuntapolitiikka, 84*(1), 61–69. www.julkari.fi/handle/10024/137610

Pulkki, J., Varpanen, J., & Mullen, J. (2021). Ecosocial philosophy of education: Ecologizing the opinionated self. *Studies in Philosophy and Education, 40*(4), 347–364. https://doi.org/10.1007/s11217-020-09748-3

Rönkä, M. (2020). Kotimme on tulessa. Ympäristöaiheet ovat jo valtavirtaa myös lastenkirjoissa. *Onnimanni, 3,* 4–8. https://lastenkirjainstituutti.fi/2016/wp-content/uploads/2021/03/Onnimanni-2020.pdf

Saarenmaa, K. (2021). Zoomerit muuttuvassa mediamaailmassa. In R. Hanifi, J. Haaramo & K. Saarenmaa (Eds.), *Mitä kuuluu vapaa-aikaan? Tutkimus, tieto ja tulkinnat* (pp. 133–157). Tilastokeskus. https://urn.fi/URN:ISBN:978-952-244-683-1

Saari, A., Tervasmäki, T., & Värri, V. (2017). Opetussuunnitelma valtiollisen yhtenäisyyden rakentajana. In T. Autio, L. Hakala, & T. Kujala (Eds.), *Opetussuunnitelmatutkimus. Keskustelunavauksia suomalaiseen kouluun ja opettajankoulutukseen* (pp. 83–108). Tampere University Press. https://trepo.tuni.fi/bitstream/handle/10024/102624/Autio_ym_Opetussuunnitelmatutkimus.pdf?sequence=1&isAllowed=y

Salonen, A. (2014). Ekososiaalinen hyvinvointiparadigma – yhteiskunnallisen ajattelun ja toiminnan uusi suunta täyttyvällä maapallolla. *Sosiaalipedagogiikka, 15,* 32–62. https://doi.org/10.30675/sa.122634

Schneider-Mayerson, M., Weik von Mossner, A., Małecki, W. P., & Hakemulder, F. (2023). Introduction: Toward an integrated approach to environmental narratives and social change. In M. Schneider-Mayerson, A. Weik von Mossner, W. P. Małeck & F. Hakemulder (Eds.), *Empirical ecocriticism. Environmental narratives for social change* (pp 1–30). University of Minnesota Press.

Seutu, K. (2020). Ein Überblick über die Klimathematik in der Kinder- und Judendliteratur. In I. Schellbach-Kopra, G. Schrey-Vasara, S. Moster, & S. Grünthal (Eds.), *Jahrbuch für finnisch-deutsche Literaturbeziehungen* (pp. 49–54, Nr. 52). Deutsche Bibliothek.

Siljander, P. (2014). *Systemaattinen johdatus kasvatustieteeseen.* Vastapaino.

Sintonen, S., & Kumpulainen, K. (2018). Monilukutaito moninaisuutena, toimintana ja osallisuutena. In R. Ruuskanen (Ed.), *Mitä tarkoittaa?: Mediataide monilukutaidon lähteenä* (pp. 6–12). AV-arkki.

Turp, M. (2023). Imaginative engagement with the climate crisis: The role of climate and ecology fiction. In L. Aspey, C. Jackson & D. Parker (Eds.), *Holding the hope: Reviving psychological and spiritual agency in the face of climate change* (pp. 111–121). PCCS Books.

Värri, V. (2007). Kasvatusfilosofian tärkein tehtävä. *NIIN & NÄIN – Filosofinen aikakauslehti, 14*(52), 70–73. www.netn.fi/files/netn071-16.pdf

Värri, V. (2018). *Kasvatus ekokriisin aikakaudella.* Vastapaino.

Vettenranta, J., Puhakka, E., Rautopuro, J., Vainikainen, M. P., Välijärvi, J., Ahonen, A., Hautamäki, J., Hiltunen, J., Leino, K., Lähteinen, S., Nissinen, K., & Nissinen, V. (2016). PISA 15 Ensituloksia. Huipulla pudotuksesta huolimatta. *Opetus- ja kulttuuriministeriön julkaisuja, 41.* http://julkaisut.valtioneuvosto.fi/bitstream/handle/10024/79052/okm41.pdf

Whitlock, J. (2023). Climate change anxiety in young people. *Nature Mental Health, 1,* 297–298. https://doi.org/10.1038/s44220-023-00059-3

Wolf, M., & Barzillai, M. (2009). The importance of deep reading. *Educational Leadership, 66*(6), 32–37. http://doi.org/10.1.1.461.7284&rep=rep1&type=pdf

Wong, C. C., & Kumpulainen, K. (2019). Multiliteracies pedagogy promoting young children's ecological literacy on climate change. In K. Kumpulainen & J. Sefton-Green (Eds.), *Multiliteracies and early years innovation* (pp. 95–114). Routledge. https://doi.org/10.4324/9780429432668

11 Moving Forward

Fostering the Growth of Climate-Resilient Students

Erin Madon and Astrid Steele

Positionality Statement

This story is Erin's story – it is told by her; Astrid's role, as second author, was to provide many of the research references and perspectives that support and give weight to Erin's experiences and understandings.

A chapter, presented as a case study addressing climate anxiety in the classroom, would be incomplete without first exploring my own background and perspectives. My lens on climate change, and the emotions attached to it undoubtedly affect the way I teach science (Lombardi & Sinatra, 2013; Ojala, 2023). I grew up in Thunder Bay, a mid-sized city in the northern portion of the province of Ontario, Canada, surrounded by nature, on the tip of the largest lake in the world. I was raised by liberal parents and had a grandparent who strongly advocated for sustainability.

Growing up, my family lived below the poverty line, which has given me strong convictions on the need for equality and equal opportunity. That said, as an adult, I live differently. I am part of a higher economic bracket and acknowledge that as a Westerner, my lifestyle is a huge contributor to climate change. I am afforded the time to think about this issue, because I live in a home where I do not worry about my financial or emotional stability, which cannot be said for all of my students.

I teach in a large urban centre in Ontario and my students are extremely diverse. They come from all over the globe and have lived experiences very different from my own. I have taught students from countries already devastated by climate change – they may be fleeing it. Conversely, I have taught some who do not have climate on the radar. Their countries of origin are in the grips of war, and many still have family there. Some have experienced extreme poverty and continue to support families at home and overseas. All these factors affect their views on climate change, and their capacity to learn and care about it.

This chapter will be a glimpse into my world as a practicing science educator, a snapshot of how climate change education and climate anxiety have manifested themselves throughout my career and my attempts to address them. As Burkhart

DOI: 10.4324/9781003494416-13

(2016) points out "As responsible educators, I suggest that we need to be doing our own inner work, and be comfortable navigating our own emotional landscape, in order to effectively hold space for students to do this as well" (p. 77).

My Experiences with the Evolution of Climate Change Education and Climate Anxiety

Denial

In 2005, when I began teaching in Thunder Bay, Ontario, many students, staff, and parents did not believe in climate change, then known as global warming. Climate education was rarely talked about in school and was not considered when planning units of study unless you were part of the environmental movement. My first high school practicum was teaching college-level biology, and my associate teacher tasked me with the Plants unit; a subject considered optional compared to "important" units such as Genetics or Microbiology. I was advised to cover the structure and function of individual plants rather than the interconnections between plants and humans. The idea of including climate change in that unit was never discussed.

In the early 2000s, there were not many official avenues to teach about climate change, and – depending on the school – environmental education was controversial. Media was beginning to cover climate change more regularly, and as the scientific community and meteorological societies released more and more data showing consensus on anthropogenic climate change, teachers began talking about it. When Al Gore's movie *An Inconvenient Truth* was released in 2006, companies all over the Western world began to capitalize on this new, shocking idea. For example, Roots Canada sold STOP GLOBAL WARMING recycled leather bracelets, and anyone who wore one was seen as taking a stance – climate change had become a political issue.

Teachers who showed *An Inconvenient Truth* to their classes, or who wore the Roots bracelet, quickly became unofficial spokespeople for climate change. Anyone who cared about sustainability was in the spotlight; judged on how they lived their lives. If they did not recycle, or drive a small car, they were considered hypocrites. It was an isolating time; some staff actively avoided environmentally focused teachers for fear they would be asked to change their personal lifestyles in favour of "green" consumer choices and actions (which evoked feelings of guilt), and/or to change teaching practices to place more focus on climate change and sustainability (which involved extra work). These responses echo Lombardi and Sinatra's (2013) who found that, like their students, teachers also hold emotions associated with climate change; possibly they feel threatened that they might be expected to change their practice in the face of insufficient knowledge about the science of climate change, and the lack of a skill set to convey that information to students.

I felt that if I wanted to be accepted into teacher social groups, I had to shelve conversations about climate. This was worrying because I wanted to have these meaningful conversations with my colleagues, but every time I would bring it up climate change, they would dismiss it, deny it, or steer the conversation onto lighter topics. Pihkala (2020) speaks to this lack of conversation about climate change among educators, proposing that "emotional cultures and regimes [in educational settings] usually do not endorse this kind of behaviour, not to mention the traditional power structures in educational settings" (p. 21). While it is widely suggested that students be provided with the space to share their emotions openly and safely about climate change (e.g. Baker et al., 2021; Pihkala, 2020), the same was not available to me at that time.

Simultaneously, administrators encouraged environmentally focused educators to form eco-groups with like-minded people so that the school could boast about its green initiatives without having to change behaviours or talk about it as a full staff. This was extremely frustrating because it siloed any environmental activities we implemented and stunted our capacity to truly effect school-wide change.

Indifference

The Ontario high school science curriculum was updated in 2008, marking a huge win for those of us who knew climate change was serious and immediate (Ontario Ministry of Education, 2008). In virtually every science course, the idea that humans were impacting the environment – and more importantly, that humans had a responsibility to protect it – was added to the curriculum. Although the idea of sustainability did exist in the curriculum prior to 2008, the volume with which environmental considerations appeared, and the addition of the term climate change, highlighted the importance of the issue and that it needed to be specifically addressed in schools.

The structure of the curriculum outlines, however, did not change. The new Overall Expectations still took aim at broad overarching ideas addressing interconnections between and among humans and the natural world, expectations that are difficult to assess in terms of student learning, because of their subjective nature. The Specific Expectations in the science curriculum were still numerous and content/fact oriented; they largely provided the basis and bulk for objective student assessments in the sciences. In effect, it seemed like the science curriculum had still been structured as checklists of content-based topics to be covered. I felt that if the curriculum expectations were not so numerous and specific, or if climate education were the focus of core planning, this structure could have been improved further and allowed for richer projects that provide deeper understandings and connections between units.

The response to the new curriculum among my colleagues was varied. I recall science teachers complaining that a Space unit had been added to the Grade 9 curriculum, and many could not fathom why it was worth teaching as compared to the more recognized science branches of chemistry, biology, and physics (you would be hard-pressed to find someone with this attitude in 2023). This was the same with climate change. Public opinion data presented in 2016 by Mildenberger et al. showed that 60–90% of Ontarians thought the Earth was getting warmer, yet only 50–70% believed this was solely a result of human activities. Anecdotally, it seemed many of my education colleagues no longer denied climate change was happening, rather they dismissed it. They felt that while human activity did impact the climate it could not possibly cause catastrophic damage, and many thought we had more time to solve the problem. At this stage in my career, I wanted to toe the line and was concerned that if I taught mostly about climate that I would not be preparing students for their senior science courses. Yet I was beginning to see how I might link the science units and climate change ideas.

Like my colleagues' varied responses, society was also polarizing around the topic of climate change. News outlets made it seem as though climate change was an issue to debate – rather than a fact – by giving equal time to climate sceptics and climate scientists. Political parties had to take a side. Many parents (and their children) saw climate change as a political issue rather than a scientific one and complained to me personally about including it in their education.

For these reasons, the new curriculum – and more specifically the Grade 10 Climate Change unit – was not well implemented. It was viewed as controversial, less important than the standard sciences, only to be covered if time permitted, or covered elsewhere in the curriculum. Despite the Climate Change unit being robust in nature compared to that of other Canadian provinces (Wynes & Nicholas, 2019), my colleagues at the time felt uncomfortable with the subject matter because they did not have the base knowledge to teach it, nor did they want to have to defend themselves to unhappy parents and students. In a study of over 1,000 Canadian teachers, Field et al. (2020) corroborated that only one-half to one-third of the teachers surveyed felt they had the knowledge and skills to teach about climate change, and, as Lombardi and Sinatra (2013) found, teachers who feel uncomfortable with teaching controversial issues tend to avoid them altogether or treat them very superficially. Climate as a subject was evolving faster than teachers could keep up, meaning that by the time it was being taught, the social and political climate – as well as the science and predictions about climate change – had become both more dire and more immediate.

Anxiety

Between 2010 and 2020, the media landscape surrounding climate change intensified, and an increasing number of Canadians acknowledged that climate change was driven by human actions. In 2019, a comprehensive survey of 3,196 Canadian (K-12 teachers and students, parents, and general public) conducted by Lakehead University and LSF (Learning for a Sustainable Future) showed that

86% of those surveyed believed that climate change was real and 79% were concerned about its impacts (Field et al., 2020). This belief, paired with the emergence of smartphones and Wi-Fi in schools, led to an increasingly climate-aware student population.

Every year I would poll my classes to see who did and did not believe in climate change, and if they had relatives or friends who were still unsure. By the time the Covid-19 pandemic emerged, virtually all students in my classes believed in anthropogenic climate change. Superficially, this seemed a good thing; however, this belief did not appear to translate into optimism, nor did it account for students' feelings about a changing climate, and most importantly, whether or not they thought humanity could solve this problem.

Many teachers I knew were now covering climate change in their lessons but still only in a basic way. The new rhetoric was that students were bombarded with climate talk everywhere, so it did not really need to be taught. This led to students hearing repeatedly how the planet was warming and that there were going to be grave consequences with no real solution in sight. On the surface, it seemed students, staff, and parents were experiencing climate information fatigue; they were becoming apathetic before the subject matter had even had the chance to take off. However, in my opinion, it was not apathy at all. It was a repressed growing anxiety, because if they thought about climate change and its consequences, it made them anxious. As Ojala (2023) states, "the climate-change problem at its core relates to existential emotions like anxiety, worry, and ambivalence, and one could argue that this needs to be taken into account when teaching about climate change" (p. 3). I was, and still am, anxious about our future even though I am confident we can reverse some of the worst effects of climate change; I can only imagine how those who have relied solely on the media for information feel.

According to the Canadian Psychological Association (CPA), eco-anxiety and habitual ecological worry can happen before, during, and after related climate events, and can manifest itself in many different ways (McCunn et al., 2020). With repeated warnings about catastrophic events, the CPA reports that people can feel worried for personal safety and the safety of loved ones, which can be associated with feelings of loss, helplessness, frustration, and the inability to improve the situation. I saw this not only in staff and students at school but also in family and friends, which led me to believe students were truly being bombarded with worry from all sides.

Indeed, following more discussions with students and colleagues, it seemed no one believed we could solve the problem; they could not see how anything they did would make a difference. Anthropogenic climate change and its solutions were too big, too overwhelming. The joint Lakehead/LSF study supported this observation with evidence that 46% of students who understood that climate change is caused by human actions did not believe attempts to mitigate the situation would be effective (Field et al., 2020).

As the world started to experience the more intense natural consequences scientists had predicted, anxiety grew. In my city, there were a series of extreme weather events that affected many students and school staff: tornadoes in 2018, floods in 2017 and 2019, the derecho windstorm in 2022, and the most recent freezing rainstorm in 2023 – the last two causing power outages over multiple days. With each event, I found more and more students shying away from climate discussions. The CPA states that during and after climate events, people can feel deep sadness, depression, grief, Post-traumatic Stress Disorder (PTSD), aggression, interpersonal difficulties, substance abuse, and even suicide (McCunn et al., 2020). Children specifically can experience low mood, social withdrawal and anxiety, and children with lower socio-economic status, and those with pre-existing mental health conditions are particularly vulnerable (McCunn et al., 2020). The, "we could all die," existential nature of climate change makes it an emotional topic. It is about physical survival; it has a moral dimension that questions how we should ethically live on the planet, and it has a spiritual component that questions the meaning of living and the meaning of actions taken or not taken (Ojala, 2023). Small wonder that I saw many of these symptoms growing in my students, especially social withdrawal, increased generalized anxiety, and feelings of helplessness. The students' feelings were exacerbated after the onset of the COVID-19 pandemic. Faced with long periods of isolation, an increased global worry compounded what they were already feeling. During the most restricted times, teachers were told to keep education light and positive and to address students' emotional needs as best as possible. This meant teaching less about climate change and its effects because those were topics that caused stress and worry.

After the COVID-19 restrictions eased, I sensed that students and staff were struggling mentally. Some staff shared with me that they were having trouble teaching about climate because it made them feel depressed, they felt they did not have the resources, and/or they were preoccupied with catching students up on lost learning. In general, a learned helplessness seemed to have settled into both students and staff due to the additive effect of multiple stressors over time – the pandemic, the isolation, the extreme weather events, and the continuous bad news all over the media. It seemed to me that students were coming to class with lower self-esteem, less task persistence, decreased motivation, increased anxiety, depression, and difficulty concentrating.

Empowerment and Hope

Even before the pandemic, it was clear that climate education needed to address not only the academic component but also the psychological and social emotional aspects of climate change. We needed to prepare students for the future in a way that did not make them feel sad or helpless. In such a short time, I had seen shifts from denial, to indifference, to anxiety in the classroom, and I felt we were at a tipping point – I needed to adapt my thinking and my practice when teaching about climate change in order to move forward, or else I risked a continuation of those negative emotions in my students and in myself. I believe the way forward lies in empowerment. We need to teach our students about existing solutions and what we, and they, can do to address the negative effects of climate change. We all need to feel like we are making a difference. We need to arm ourselves and our students with tools, skills, and hope.

Pedagogical Strategies Used to Counter Climate Anxiety

Anxiety – an emotion characterized by feelings of tension, worried thoughts, and physical changes like increased blood pressure (Anxiety, 2018) – is – at its heart– a psychological issue, with clear treatment options to help alleviate symptoms. But educators are not psychologists. We are trained to administer curriculum, not to provide expert care for a student's social, emotional, and physical needs. Yet teachers and education staff do this anyway. When students are hungry, we organize breakfast programmes. When students need clothes, we set up clothing swaps. When students are anxious or sad, we reach out to them and to their parents. We seek any support to which we have access because we care. Nel Noddings (1988) speaks to this ethic of care with which teachers respond to their students' needs, with an empathy that encompasses both their academic and emotional struggles; Noddings recognized that teachers and students can build important caring relationships with each other (with the teacher carrying the heavier responsibility and burden for the caring). Now more than ever, with children and teens showing up to schools anxious, depressed, in need of mental health support (Wiens et al., 2020), teaching as an act of caring for each other (Noddings, 1988) seems vital to both students' and teachers' well-being.

Although educators are not trained psychologists, there are a few well-documented best practices we can use to help address anxiety in the classroom including increased physical movement, education and awareness of the issues causing the anxiety, mindfulness, structured problem solving, and self-esteem building (Public Health Agency of Canada, 2009; Department of Health Victoria, 2012). Some students I teach have been clinically diagnosed with an anxiety disorder, whereas others express they are stressed out, mildly worried, or not anxious at all. Some of these students may have eco-anxiety, while others may have more generalized anxiety that may or may not include worries about climate. Regardless of severity and type, best practices remain the same and can be included in programme development regardless of student age or subject taught. Over time and giving attention to the increased need to teach about climate change, while also recognizing and empathizing with my students' emotional responses to their learning, I guide my teaching using four main pillars that seem to yield positive results.

Approaches to Addressing Student Anxiety

This section will describe the four strategies I use to address student climate anxiety and to do so a brief description of the structure of the Grade 9–10 science curriculum in Ontario is helpful. There are five units of study in both grades: *Careers/the Scientific Method, Biology, Chemistry, Physics,* and *Earth and Space Science*. Each unit has Overall Expectations as well as guiding *Specific Expectations* as described earlier in the chapter. Thankfully, in recent years, it is now recognized that larger projects covering one Overall Expectation may be more beneficial than "checking the boxes" of Specific Expectations, as large projects can lead to increased complexity, deeper understanding of subject matter, increased capacity to take action, and gains in critical thinking (Maros et al., 2023).

Since 2015, I have regularly taught Grade 9 and 10 science, and remaining in one subject area has given me the opportunity to delve deeply into the curriculum and reflect upon and evaluate classroom learning. I feel privileged in that the Grade 9 and 10 science courses require that I teach climate change and sustainability, so I do not have to reach far to connect climate education and other science content. However, with this sanction comes responsibility; I still feel pressure from some peers in the science community to ensure that all the traditional science content is covered, which requires finding a delicate balance between keeping the programme objective and fact-based, while allowing for emotional connection, student well-being, curiosity, and interest. In the Grade 9 and 10 courses that I teach, I have implemented four main approaches and illustrated them here using a few curated examples. As I typically have the same cohort of students in both Grades 9 and 10, these concepts are introduced in a continuous way over a two-year period.

Spiralling

First, I spiral curricula around the themes of sustainability, social justice, and climate change. By infusing climate lessons into everyday science learning tasks, students become both more cognizant and educated on associated issues, which Gu et al. (2020) report diminishes uncertainty and alleviates anxiety. When students see a topic repeatedly and how it weaves itself through all aspects of life, they are less uncertain, more educated, and more aware which in turn helps them feel empowered and capable of positive action (Baker et al., 2021; Ojala, 2015; Ojala, 2017).

The Grade 9 course spirals around the central pillar of sustainability. The Biology unit covers ecosystems, the carbon cycle, and climate change, and these concepts are woven throughout the other units to answer centrally themed questions: *How can we as humans move forward in a positive relationship with nature? How are we a part of nature, not apart from it? What does our ideal world look like?* Through a series of activities, guest speakers, and direct instruction, these ideas compound over time into deeper understandings.

It is not difficult to tie the content from the other units to the topic of sustainability. In the Chemistry unit, we learn about bioaccumulation, and the story of the Grassy Narrows First Nation (http://thirstyforjustice.ca/). This has led to passionate letter-writing campaigns and petition signing by students. We talk about which elements on the periodic table are in plentiful supply, and those that are threatened, then brainstorm alternatives. We learn about physical and chemical properties of elements and compounds, and how these properties can be used to solve current technological problems. We also discuss how mining has hurt ecosystems, and the power we have with the choices we make as consumers. For example, a discussion about diamond mining led to students discovering the existence of lab-made diamonds. They reported feeling good knowing there were solid alternatives they could seek out that did not hurt the environment.

In our Space unit, we talk about the building blocks that are needed for life on other planets, and why we are exploring other planets to live on. We discuss the amount of space debris that is in our atmosphere, and who is responsible for it. We consider future careers in engineering, space, and technology, and students feel empowered when they see how scientists are tackling many of the world's sustainability challenges.

In the Electricity unit, we visit https://gridwatch.ca/ to examine how and where power generating stations are powering our homes and at what percentages. We investigate the longevity of each station; we talk about renewable versus non-renewable energy. Students complete a project on electrical grid management, which addresses literacy and numeracy, solutions-focused learning, empowerment, and action – now and in the future. Students become confident in reading news articles about Ontario's energy grid, because they feel like they have solid knowledge about how it works and its future direction.

The Grade 10 programme is taught in the same way. The Climate Change unit is always studied first, which then frames the learning in our other units. We develop our thematic questions: *What is the science behind climate change? Who causes it? Who is affected by it most? What is our responsibility?* And most importantly: *How do we solve it?*

The largest unit covered in Grade 10 is Chemistry, which can also include a climate focus. When studying molecular compounds, we learn about how the shape of greenhouse gas molecules traps heat. When learning about chemical reactions, we link combustion to climate change. When tackling acids and bases, we discuss the process of ocean acidification, and how this leads to the destruction of coral reefs. In a project, students research a group, government, or institution working on solutions in any of the previous domains. This is sorely needed information: as the LSF study showed that 43% of Canadians failed a basic knowledge test on climate change, though they reported feeling confident in their answers. Only 14% of respondents correctly answered eight or more questions out of ten (Field et al., 2020).

Spiralling continues in the Biology unit, when we learn about cells, homeostasis, and the organization of our universe. We discuss cultured meat, fermentation, genetic modification, and how these solutions apply to future farming. We also compare and contrast animal and plant systems and discuss how plants are the unsung heroes of our planet. Then, in the Optics unit we learn about the electromagnetic spectrum, and how light is absorbed, reflected, and refracted. We talk about feedback loops, albedo, and the effect of less snow at the planetary poles on global temperature. By the end of Grade 10, the students leave with a solid knowledge of climate change and its possible solutions, which goes a long way to alleviate anxious feelings and eco-anxiety (Ojala, 2015).

Outdoor Learning

The second strategy is to include outdoor learning as part of lesson planning, which provides benefits for the climate-anxious student. Outdoor learning has many positive effects such as improving mindfulness, bolstering self-esteem, allowing students to take ownership of natural spaces, increasing movement, increasing engagement, and encouraging collaboration (Bjorge et al., 2017; Pihkala, 2020), each of which is directly tied to decreased anxiety (Curll et al., 2022). One of the summative activities we do in Grade 9 is to track the life of a native species over the course of a semester. This brings learning outside both at school and at home. We first go for walks around the school neighbourhood to discover the plants and animals who share this space with us. No matter the weather or the students' moods prior to a walk, I always notice a huge difference after we go outside. The students talk to each other more, they smile more, move more. At the end of the class, they leave happy, and they ask to go outside again. As we continue the project, I ask students to pick a species that speaks to them, then they draw it, take pictures of it, and explore it where they themselves live. Some choose to study a tree they have grown up with in their backyards. Some look to the skies and watch the birds fly by. Others are intrigued by how many mosses, lichens, and mushrooms they see, and choose a species they can touch and interact with. By the time we go outside for a third or fourth time, the students are focused, pensive, and more willing to take risks (such as trying to jump across the rocks from one side of our stream to the other). When the students are outside, they are visibly less anxious.

The final products from this project yield some of the deepest learning and highest marks the students achieve in the course. The students create posters, videos, journals, photographs, and models of their species. It is truly one of my favourite projects because the joy students had in learning is visible in their work. And to this day, current and prior students find me to show me photos of where they continue to see their chosen species.

Growth Mindset

Harvard Business School Online contends that a growth mindset – the idea that ability and intelligence can be achieved through effort – leads to adaptability, resilience, humility, and the capacity to take constructive criticism and use it to make the best final products (Cote, 2022). I try to foster a growth mindset in my students through constructive feedback, project-based learning that continually evolves and iterates, and providing opportunities for positive action; these are intended to develop a student's sense of personal control. Focus on a growth mindset approach has been shown to decrease generalized anxiety in adolescents, which may translate to decreased climate anxiety as well (Schleider & Weisz, 2017). During the pandemic, many students shared that they were anxious, and that their overall mental health was not good. Having already seen the decline in student well-being over a period of years due to climate change and other factors, this increased anxiety was troubling. At this point, many teachers looked for guidance on how to develop resilience, build self-esteem, and help alleviate anxiety in our students. This led some of us to develop a growth mindset in our classrooms through positive reinforcement, growth messaging, and increased feedback/opportunities to improve.

In my learning spaces, students are greeted with a big smile at the door. There are messages around the room that foster positive self-talk and resiliency such as "FAIL- First Attempt In Learning" and "I don't know how to do this – yet." I mark less but assign richer tasks, then allow students to edit, and improve their work multiple times prior to marking. As a goal, I personally try to have three positive interactions and/or give three positive comments before giving constructive feedback. I also bring in mental health speakers who suggest age-appropriate techniques (meditating, journaling, having a good cry, getting enough sleep, etc.) to help students develop self-esteem, improve mindfulness, and ease anxiety. Since using these strategies, I have noticed in some students increased confidence, less fragility, and an overall decrease in worry during class time. The hope is that by developing resilience and fostering positivity and growth, this will help students handle their anxiety – climate based or otherwise.

Solutions-Focused Approach

Finally, by employing a solutions-focused approach, students can become confident, resilient, critical thinkers (Redman, 2013), equipped with tangible solutions to climate concerns at every level – from the individual to the global community. I find this strategy is the most effective at countering climate anxiety. When students look at a problem and can find solutions, I see their empowerment growing. They appear less worried and more hopeful. The example I share here is based on the Grade 10 Climate Change unit.

After learning about the causes and consequences of climate change, students pick a country in the world that interests them. We use a student-developed chart to fill in important climate-related information about the country of their choice (e.g. total emissions, emissions per person, average temperature, major industries, climate action plans, vulnerability scores). Students evaluate their chosen countries on two measures: how much they contribute to climate change, and how much they are affected by it, then students visually report this with colour posters:

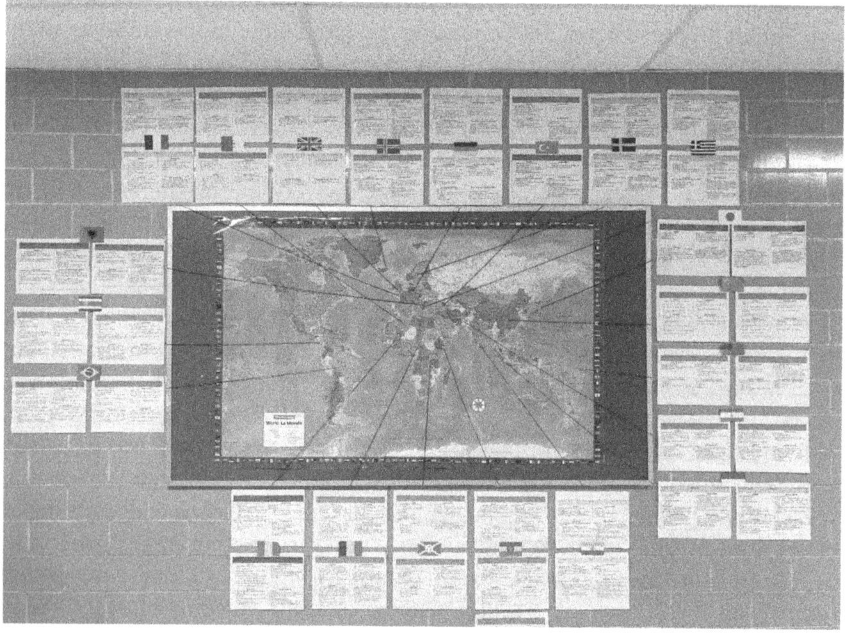

Figure 11.1 Climate Change: Cause and Effect Project

The project also requires students to evaluate whether their chosen country's plans to address climate change are substantive enough to achieve the global goal of reducing warming to 1.5°C. This allows students not only to consider a more global perspective but also to make connections between curriculum content and real-world scenarios. They see that the countries most affected by climate change are often the ones least responsible for it. The project helps students develop critical thinking skills and an appreciation for nuance when it comes to global solutions. The power of this project is seeing how climate change can be tackled in different ways, and students hopefully begin to feel less worried knowing that governments around the world are working on many different solutions.

To be less climate anxious, students need to believe that climate change can be mitigated and that they can make a positive contribution (Baker et al., 2021; Curll et al., 2022; Ojala, 2015; Pihkala, 2020). A second Climate Change unit project focuses on finding solutions to climate change. Using data collected on the overall impact of diverse solutions (https://drawdown.org/solutions/table-of-solutions), students are tasked to remove 500 gigatonnes of carbon dioxide equivalents by the year 2050 to stop the world from warming more than 1.5°C. Student approaches to this project are innovative and creative. Some choose to tackle small changes over multiple sectors (such as energy, transportation, and education), while others choose to concentrate efforts by focusing on the highest-yielding solutions (onshore wind turbines, utility-scale photovoltaics, and food waste management, for example) and fully develop the solutions across countries. Some students enjoy learning about the more obscure solutions that are not always well understood or talked about in the media, which lead to fruitful discussions on topics such as refrigerant management, or the education of women and girls. Many students have reported this activity has helped them feel less worried, and they are surprised at how possible the solutions are, which gives both them and me hope.

The strategies presented by the students can be translated into other classroom settings or subjects. For example, in foods classes, students can learn about composting, recycling, and packaging. I have seen teachers in other subject areas using the United Nations Sustainable Development Goals (https://sdgs.un.org/goals) to design lessons that infuse social responsibility and climate action. I believe the point is to include climate change in diverse classroom conversations and to frame it in a positive way.

Student Feedback of the Programme

Student feedback is imperative not only when designing curriculum but also when evaluating success in the classroom. To understand what students were getting out of Grade 9 and 10 science programmes, I developed an informal survey that asked (on a scale of 10):

- What was your level of knowledge about climate change and sustainability practices before and after these courses?
- What was your level of climate anxiety before and after these courses?
- What was your level of belief that climate change is solvable, and your confidence in your capacity to personally help solve them?

I also asked students to provide additional written feedback. The response rate on the voluntary survey was quite low (30%), with a total sample size of 8. This section includes a summary of the survey data (see Table 11.1), as well as general feedback I received from students.

Table 11.1 Student Survey Responses

Student	Level of Knowledge			Level of Anxiety			Level of Confidence in Solutions		
	before	*after*	*+/−*	*before*	*after*	*+/−*	*before*	*after*	*+/-*
A	6	9	+ 3	8	4	− 4	0	10	+ 10
B	4	8	+ 4	7	5	− 2	2	10	+ 8
C	4	8	+ 4	8	5	− 3	5	8	+ 3
D	6	8	+ 2	3	7	+ 4	3	9.5	+ 6.5
E	4	8	+ 4	3	7	+ 4	4	7	+ 3
F	4	10	+ 6	3	9	+ 6	7.5	10	+2.5
G	3	9	+ 6	4	9	+ 5	3	9.5	+ 6.5
H	5	9	+ 4	5	8	+ 3	7.5	6.5	− 1
	4.1 point av. increase in level of knowledge			1.6 point av. increase in level of anxiety			4.8 point av. increase in level of confidence in solutions		

Recognizing the low number of survey participants is limiting in terms of suggesting general trends; nonetheless, some interesting results emerged. As indicated in Table 11.1, the responses to Level of Knowledge indicate that all the students felt that they had gained in that regard with an average increase of 4.1 points, indicating that even though students are bombarded by climate messaging, the actual breadth of their scientific knowledge is sparse. Students commented that Climate Change unit was their favourite unit of study, that they learned it more deeply, and that it was engaging. Student A reported: "Before this course, I believed I had all the knowledge I needed, yet after this course I realized how much more there is to learn."

Similar results are found in the Level of Confidence in Solutions, with an average increase of 4.8 points and only one student reporting a drop in Level of Confidence. Expecting that an increase in knowledge and confidence in finding solutions should lead to a decrease in Level of Anxiety, I found the opposite was true. Three respondents felt less anxious after the course, but five felt more anxious. The less anxious group reported that learning about the many solutions and what is being done already made them feel better. The more anxious group responded that they did not realize the seriousness of the issue until they learned more about climate change in depth, and how much was happening already. One student reported feeling anxious because even though the answers are there, humanity is not doing enough. This was tough to see as an educator because I feel the same way. The best I can do for this student is to continue to teach with hope, and to follow up with good news stories. It is difficult to strike the right balance: if we do not teach about what is truly happening, students may feel less anxious in the short term but more anxious in the long term. They may feel ill equipped and angry when faced with the more devastating effects of climate change in the future. Indeed, in their far-reaching survey of Canadian students, Galway and Field (2023) found that students want to learn about solutions to mitigate climate change – they want increased instruction and support as they face their futures.

The finding in my survey that the more the students learned about climate change the more worried they became, but also more confident, implies that their self-efficacy, that is, their personal belief in their capacity to act to achieve a goal, had increased. Although they were more worried, they believed that they were more capable of addressing climate change. An increase in self-efficacy also suggests that the students would be less likely to experience negative effects from their worry. These results from my study mirror those reported by Baker et al. (2021).

Students' comments regarding their understanding of solutions were very positive. Student G wrote, "After learning about the variety of things people are doing and the options that I have to be a part of them, I feel more confident in the impact I can have."

Student A said:

I have realized how possible fixing climate change is, before this course I always believed it was too late and that we were doomed! But after this class I understand that solutions exist and although it may be challenging, I'm now sure the human race can overcome this problem.

Overall, the students felt more capable of looking at the issue on a national and international level and reported feeling much more confident.

Student B wrote:

Before grade 9 we were taught more about the dangers of climate change, never really the fact that we actually can do something about it. . . . Ms. Madon showed a lot of confidence in our community and us as students in being able to work on climate change, which is very inspiring and reassuring.

Successes and Challenges

General comments about my students' science learning and how they feel in my classroom have been overwhelmingly positive, whether they come from staff, parents, or the students themselves. I am not the only climate educator at my school; staff members have shared the wonderful things they are doing in their classrooms, which has been a gift that has allowed me to refine and add to my own programme.

Former students come back to visit on a regular basis, and they want to talk about climate change – their fears, their ideas, and even just what they are doing outdoors. They send me pictures of their tomato plants and tell me they plan to follow careers with an environmental focus. They come to look at the worms and pick the lettuce in our growing tower. We share a bond forged in nature and connection – one that is not soon forgotten. These moments with my current and former students echo Nel Noddings' (1988) charge to create with our students' caring relationships as we teach and learn together. These gestures not only warm my heart but also give me hope that students are feeling more positive after leaving the junior sciences.

A former student has shared:

> The 9 and 10 climate units were what I would consider as the most important and interesting units that I've learned so far in high school. These units completely changed my view on the importance of climate change education and climate solutions. I now think about my impact on climate change multiple times a day. I continue to think about what I learned while trying to implement positive climate practices in my day-to-day life.

In my corner of the world, there will always be obstacles when it comes to teaching about climate change; however, we are making strides. There are now many initiatives led by students and staff that tackle both climate and social justice; students are learning about and leading climate action in our schools and are feeling empowered by it. Our students are benefiting from the inclusion of Indigenous ways of knowing in the curriculum through guest speakers, courses dedicated to Indigenous authors, and activities using traditional knowledge.

There is a growing movement towards outdoor learning and returning to nature, for example, our school is piloting a course in *Green Industries* that centres around our courtyard, allowing students to learn about gardening, landscaping, invasive species removal, flora and fauna identification, green tech, and green jobs. This programme and others like it breathe life into the school with plants, growing towers, and garden beds in classrooms and on the school grounds. These all have a positive and calming effect on both students and staff, who have been inspired to bring nature into our spaces.

And there are still definite shortcomings in our school system. As discussed in the first section, climate education is often relegated to a small, dedicated group of teachers. Although most schools likely care about the environment, funding issues, lack of knowledge, difficulty in applying true and substantive changes, time constraints, and lack of political will have led to a lot of talk about climate change without a lot of action. Teachers are often left to attempt to implement change without coordinated oversight. Thus, many teachers are still reluctant to incorporate climate into their lessons as they may feel underqualified, lack the time to address the topic, or are personally overwhelmed by climate change. These concerns could be addressed with professional development on climate education offered to all teachers instead of only science or climate educators, a recommendation shared with the results of the Galway and Field survey (2023).

Final Thoughts

Despite the challenges faced in school systems and the immediate threat of climate change, there is still a tremendous amount of hope. Our children are the future stewards of our planet – some already take this role very seriously. No matter what I teach, or whom, I see hints of brilliance, resilience, and innovation. I see a brighter tomorrow. And although climate change education has had to evolve quickly and students are now faced with more anxiety and uncertainty than ever before, it will be the slow, steady gains and the small, everyday moments of learning with care and hope, that will lead us forward.

References

Anxiety. (2018, April 19). *American Psychological Association.* www.apa.org/topics/anxiety#:~:text=Anxiety%20is%20an%20emotion%20characterized,changes%20like%20increased%20blood%20pressure

Baker, C., Clayton, S., & Bragg, E. (2021). Educating for resilience: Parent and teacher perceptions of children's emotional needs in response to climate change. *Environmental Education Research, 27*(5), 687–705. https://doi.org/10.1080/13504622.2020.1828288

Bjorge, S., Hannah, T., Rekstad, P., & Pauly, T. (2017). The behavioral effects of learning outdoors. *Masters of Arts in Education Action Research Papers, 232.* https://sophia.stkate.edu/maed/232/

Burkhart, J. (2016). Singing the spaces: Artful approaches to navigating the emotional landscape in environmental education. *Canadian Journal of Environmental Education, 21*, 72–88. https://files.eric.ed.gov/fulltext/EJ1151865.pdf

Cote, C. (2022, March 10). *Growth Mindset vs. Fixed Mindset: What's the difference?* Harvard Business School Online. https://online.hbs.edu/blog/post/growth-mindset-vs-fixed-mindset

Curll, S. L., Stanley, S. K., Brown, P. M., & O'Brien, L. V. (2022). Nature connectedness in the climate change context: Implications for climate action and mental health. *Translational Issues in Psychological Science, 8*(4), 448–460. https://doi.org/10.1037/tps0000329

Department of Health Victoria. (2012). *Managing and treating anxiety.* Vic.gov.au. www.betterhealth.vic.gov.au/health/conditionsandtreatments/anxiety-treatment-options

Field, E., Schwartzberg, P., Berger, P., & Gawron, S. (2020). Climate change education in the Canadian classroom: Perspectives, teaching practice, and possibilities. *Education Canada.* www.edcan.ca/articles/climate-change-education-canada/

Galway, L. P., & Field, E. (2023) Climate emotions and anxiety among young people in Canada: A national survey and call to action. *The Journal of Climate Change and Health, 9*, 1–8. https://doi.org/10.1016/j.joclim.2023.100204

Gu, Y., Gu, S., Lei, Y., & Li, H. (2020). From uncertainty to anxiety: How uncertainty fuels anxiety in a process mediated by intolerance of uncertainty. *Neural Plasticity, 2020*, 1–8. https://doi.org/10.1155/2020/8866386

Lombardi, D., & Sinatra, G. M. (2013). Emotions about teaching about human-induced climate change. *International Journal of Science Education, 35*(1), 167–191. https://doi.org/10.1080/09500693.2012.738372

Maros, M., Korenkova, M., Fila, M., Levicky, M., Schoberova, M. (2023). Project-based learning and its effectiveness: Evidence from Slovakia. *Interactive Learning Environments, 31*(7), 4147–4155. https://doi.org/10.1080/10494820.2021.1954036

McCunn, L., Gifford, R., & Bjornson, A. (2020, December 1). *"Psychology works" fact sheet: Climate change and anxiety – CPA.* https://cpa.ca/psychologyfactsheets/; https://cpa.ca/docs/File/Publications/FactSheets/FS_Climate_Change_and_Anxiety-EN.pdf

Mildenberger, M., Howe, P., Lachapelle, E., Stokes, L., Marlon, J., & Gravelle, T. (2016). The distribution of climate change public opinion in Canada. *PLoS ONE, 11*(8), e0159774. https://doi.org/10.1371/journal.pone.0159774.

Noddings, N. (1988). An ethic of caring and its implications for instructional arrangements. *American Journal of Education, 96*(2), 215–230. www.jstor.org/stable/1085252

Ojala, M. (2015). Hope in the face of climate change: Associations with environmental engagement and student perceptions of teachers' emotion communication style and future orientation. *The Journal of Environmental Education, 46*(3), 133–148. https://doi.org/10.1080/00958964.2015.1021662

Ojala, M. (2017). Hope and anticipation in education for a sustainable future. *Futures, 94,* 76–84. https://doi.org/10.1016/j.futures.2016.10.004

Ojala, M. (2023). Climate-change education and critical emotional awareness (CEA): Implications for teacher education. *Educational Philosophy and Theory, 55*(10), 1109–1120. https://doi.org/10.1080/00131857.2022.2081150

Ontario Ministry of Education. (2008). *The ontario curriculum grades 9 and 10: Science.* Queen's Printer for Ontario. www.dcp.edu.gov.on.ca/en/curriculum/secondary-science

Pihkala, P. (2020). Eco-anxiety and environmental education. *Sustainability, 12*(23), 10149.

Public Health Agency of Canada. (2009). Mental health – Anxiety disorders. *It's Your Health.* www.canada.ca/en/health-canada/services/healthy-living/your-health/diseases/mental-health-anxiety-disorders.html

Redman, E. (2013). Advancing educational pedagogy for sustainability: Developing and implementing programs to transform behaviors. *International Journal of Environmental & Science Education, 3*(8). https://keep.lib.asu.edu/_flysystem/fedora/c258/IJESE_v8n1_Erin_Redman1.pdf

Schleider, J., & Weisz, J. (2017). A single-session growth mindset intervention for adolescent anxiety and depression: 9-month outcomes of a randomized trial. *Journal of Child Psychology and Psychiatry, 59*(2), 160–170. https://doi.org/10.1111/jcpp.12811

Wiens, K., Bhattarai, A., Pedram, P., Dores, A., Williams, J., Bulloch, A., & Patten, S. (2020). A growing need for youth mental health services in Canada: Examining trends in youth mental health from 2011 to 2018. *Epidemiology and Psychiatric Sciences, 29,* E115. https://doi.org/10.1017/S2045796020000281

Wynes, N., & Nicholas, K. A. (2019). Climate science curricula in Canadian secondary schools focus on human warming, not scientific consensus, impacts or solutions. *PLoS ONE, 14*(7), e0218305. https://doi.org/10.1371/journal.pone.0218305

12 Nature Tales and Tails

Navigating Climate Anxiety In and Out of the Classroom

Caroline Hickman

Positionality Statement

As an educator, therapist, and board member of the Climate Psychology Alliance, my work centres on integrating depth psychology with the climate and ecological crises. In my view, addressing the climate crisis requires more than practical and technical solutions; we must also embrace our vulnerability and confront painful truths, collective denial, grief, and loss. Sustainable solutions emerge from this emotional engagement, leading to informed actions. Therefore, my approach emphasizes the importance of "inner" activism alongside "outer" activism, advocating for emotionally informed, sustainable actions to address our current crises. This chapter reflects my commitment to exploring these dimensions, offering a psycho-educational framework that combines relational, psychological, and artistic methods to help young people navigate climate anxiety in and outside the classroom.

Introduction

There is a growing understanding of the mental health impacts, including distress, confusion, and anxiety that arise from increased awareness of the climate and biodiversity crisis (Pihkala, 2020; Ogunbode et al., 2022; Wray, 2022; Weintrobe, 2021; Verlie, 2022) with increasing concern about how the crisis is affecting children and young people (Griffiths, 2013; Lawton, 2019; Hickman, 2019, 2020, 2024; Hickman et al., 2021; Davenport, 2017; Marks & Hickman, 2023). To understand the impact on mental health, we need to look at the impact of direct exposure to traumatic events (wildfires, smoke from wildfires that can travel considerable distances, river and coastal floods, storms, and extreme heat) as well as indirect adverse experiences observed by witnessing the harm being caused to others including news reports showing animals and people fleeing wildfires, struggling with flooding, or listening to stories told by survivors of traumatic events (Lawrance et al., 2021; Obradovich et al., 2018). A global survey of 10,000 children and young people (Hickman et al., 2021) found that 67%

DOI: 10.4324/9781003494416-14

reported that climate change made them feel sad and afraid, with 62% feeling anxious, 57% angry, 56% powerless, and 51% helpless. The cognitive impacts were striking, with three-quarters or more saying that the future was frightening, the worldwide average figure was 75%, with young people in the Philippines at 92%, Brazil 86%, Portugal 81%, India 80%, Australia 76%, France 74%, UK 73%, Nigeria 70%, US 68%, and Finland 56%, and over half (56%) thinking that humanity was doomed.

Research has highlighted the relational aspect of this distress showing that children and young people experience emotional and mental upset due to climate change, through both direct and indirect, short-term and long-term episodes. Mental health is affected by learning about climate change in a factual way, witnessing and experiencing its impact, hearing about its impact on others, and fearing for an increasingly uncertain future (Hickman et al., 2021). Children and young people are collectively suffering from trauma, prolonged psychological and physical stress because of climate change; they have symptoms of depression, grief, and anxiety, and relationally they are feeling betrayed, abandoned, and dismissed by the people in power and governments who they hoped would, or expected to protect them (UNICEF, 2021; Hickman et al., 2021). While not seen as a mental illness, emotional and mental distress is increasingly recognized as a moral injury, a mental health relational trauma that warps and distorts a child's security in the world and trust in other people, especially those in power who they may have believed in. It involves an abuse of power, an injustice, a violation of "what's right" (Weintrobe, 2021) felt as an emotionally and physically violent act that is then minimized, dismissed, or lied about. Emotionally it is often felt as shame, anger, and disgust because of the broken trust and boundary violations. When moral rules are broken, and then lied about, a child is left feeling not just brokenhearted and scared but also despairing at human nature, no longer knowing how to feel safe in the world.

I would strongly argue that eco-anxiety is an emotionally healthy and congruent response to environmental crises (Marks & Hickman, 2023; Weintrobe, 2021) and feeling moral injury is a sign of mental health as "It means one's conscience is alive" (Weintrobe, 2021, p. 241). There remain contrasting and competing theories surrounding the nature and treatment of climate/eco-anxiety which are argued to be pre-traumatic or post-traumatic stress, collective trauma, intergenerational trauma, and an "adverse childhood experience" (Nelson et al., 2020). While there is a general recognition from professional bodies that eco-anxiety should not be diagnosed or pathologized as an illness (RCPsych, 2023), there are also attempts to categorize and differentiate between "normal" and "abnormal" forms of eco-distress as well as examinations of whether pre-existing mental health difficulties impact climate anxiety (Hickman, 2024). Because of the emergent status of eco-distress, these debates and our understandings need to continue to develop. There are also important "frames" to consider such as the different nature and degree of distress felt by children and adults, differences in experience between people in the Global North and Global South, individual vulnerabilities and stressors, the impact of pre-existing mental health vulnerabilities, and capacity to act in the face of emotional distress through activist groups, "green" projects, and ecologically oriented therapies and projects.

Counselling and psychotherapy are increasingly addressing these issues by developing "climate-aware therapy" models in clinical practice (Davenport, 2017; Mathers, 2021; Hollway et al., 2022; Anderson et al., 2024) and through groups such as the Climate Psychology Alliance. Climate-aware therapeutic models argue that rather than diagnosing and pathologizing eco-anxiety, we should support people in making sense of their distress, finding meaning in their changing world, developing community and collective approaches for support, challenging climate denial, and embracing the range of emotions associated with this experience, including grief, despair, depression, radical hope, and empathy. A depth psychology approach that explores both conscious and unconscious processes, archetypal symbolism, and the meaning of emotions (Hollis, 1996; Hoggett, 2019; Weintrobe, 2013; Lertzman, 2019; Randall, 2019) would argue that it is this very descent, depression, experience of grief, and loss that gives meaning to the experience of waking up to the climate crisis. Many children and young people I have worked with are adamant that they do not want their climate anxiety or sadness about climate change removed or treated. They argue that these feelings make sense to them and while the feelings can be distressing, at the same time they help them to make sense of the world. Their distress is made worse when they feel surrounded by others who do not seem to care, including family or friends. In practice, it is important to validate young people's feelings as meaningful, and to use a psycho-educational approach that helps them articulate and manage their feelings and thoughts.

Romanyshyn (2013) suggested that researchers should "play" with the imaginal landscape of the work. The idea of research as a climate landscape has been a metaphor or image that has been present throughout my research in the field of climate change with children in schools. I used a range of creative research methodologies including art, theatre, personification, and storytelling. We have created seagull puppets with children and then interviewed the puppets about climate change, spray-painted (organic chalk paint) a baby humpback whale onto a school playing field and then planted 2,500 crocus bulbs into the outline to create stories with the children about climate change. The stories examined the impact of warming seas on baby whales, biodiversity loss, and extinction. I worked with youth theatre groups and had 60 children enacting climate change; I have stood in grass fields, cliff tops, and woods talking with children, and collected hundreds of stories and drawings (Tidsall et al., 2009). Some of these projects are detailed later.

Combining a depth psychology approach with education has been effective in helping children develop the emotional literacy and resilience to talk about their feelings and thoughts about climate change without causing further upset or trauma. My experience spans a range of age groups and different cultural groups and has found several models that have been effective, creative, and sustainable and can be included in the curriculum in schools in arts, languages, maths, geography, music, and drama classes.

To understand the relational nature of climate anxiety, we need to be curious, use deep listening, to show respect to all forms of emotional expression (Hickman, 2020; Pihkala, 2020, 2022; Weintrobe, 2021). We need to show humility to feelings (Hollis, 1996) and to develop climate-aware psychological and educational models that can help us navigate these unprecedented challenges, both internal and external in the world today. The challenge to exploring climate anxiety in the classroom is to find ways to support children and young people to talk about climate change honestly, but without minimizing their feelings, invalidating their thoughts about the future, or further traumatizing them.

Listening to and Researching Climate Change With Children

The school projects described later seek to elicit children's feelings through listening to their voices about climate change sensitively, ethically, and meaningfully. Listening to and interpreting children's voices will always be framed by the listener and researcher as an adult, and as such the main concern is to represent the children's thoughts and feelings as accurately as possible. I have considered how to listen to children's feelings and thoughts about climate change from several perspectives, drawing on literature from social work, eco-psychology, art, psychotherapy, education, and research with children. McLeod (2008) points to the issue of "listening" being understood differently by adults and children and so to listen to children more accurately we need to learn to listen with an attitude of respect for the other, to listen and frame responses in terms of action, and to listen through an empathetic relationship. Hohti and Karlsson (2014) discuss the importance of the child having control over their own narrative using story crafting methods and the importance of reflexivity in the researcher, paying attention to how we listen and select voices in order to try to represent the "multiplicity" of the child's voice; they point to the importance of asking uncomfortable questions not just of the data but also of ourselves as adults so we do not impose too much of our own narrative over that of the children.

Nature Tales and Tails Project – Grow Your Own Whale

Funded by a public engagement grant from the University of Bath in 2016 with Denise Lengyel (University of Bath) and Sonia Shomalzadeh (artist), we designed a three-month research project to engage a mixed age group of primary schoolchildren (ages 6–11) in a project exploring their thoughts and feelings to see if we could use creative methods to talk with them about the vulnerability of children and animals in facing climate change. Using a range of creative methods, we called the project "Nature Tales and Tails, Grow your own Whale." We started by building a story arc to engage the children in a relationship with whales:

Once upon a time a Humpback whale calf appeared on your school playing field. He came to teach the children about "sand whales" and "fluking"; diving and singing underwater; and how his tail sticks out of the water when he dives deep down into the blue to feed. He told wonderful tales of other whales and the sea to them; he taught them about climate change and how whales and the oceans need our protection.

The children listened and learnt from his stories, drew many wonderful pictures of whales for him; and in return they told him stories about their lives on land and their dreams of the sea; and then they learnt that the story they were part of could sometimes be called research.

We ran three workshops with the children; all the data were collected through drawings of whales that the children completed at the end of each workshop. We then reflected on and analysed the drawings looking to see how they changed following each workshop:

In the first workshop, we started by educating the children about different whales, describing their habitats and behaviour. We introduced them to the drawings by Sonia who draws life-sized whales on sand beaches using a stick, describing them as, "drawings that last less than a day" as they are wiped out by the returning tides. This allowed us to begin to introduce themes of impermanence and vulnerability to the children, but through art rather than talking about the whales as creatures. We gave the children templates of whales to colour in the first workshop, and their drawings were colourful but did not suggest they were concerned about the survival of the whales.

We then introduced scientific information to the children, including whale calf facts and figures focused on Blue Whales and Humpback Whales. We wanted to start to create a closer relationship between the children and different whale species. We chose to focus on calves as these might be more relatable for younger children, and from a practical point of view we were about to draw a life-sized whale on their school playing field and a calf was much more manageable than a full-sized adult whale! We showed the children comparative sizes of adult whales compared with a human and a bus, talked about how much milk a baby whale drank in a day, their weight, and how long they stayed with their mother before independence.

The second workshop began with Sonia drawing a baby Humpback Whale on their school playing field (using organic chalk paint) which we then planted with 2,500 crocus bulbs (see Figure 12.1). There would then be a period of a month before the bulbs grew. The remainder of this workshop introduced the children to climate change impact on whales. We talked about warming oceans, melting sea

Figure 12.1 The Crocus Whale

ice reducing the salinity of the water, sea level rise, impact on feeding grounds, changes in migration patterns, the relationship between arctic sea ice and algae and krill and fish, shipping, and noise in the oceans. We created storyboards with the children telling the story of their workshops and collected more drawings of whales as data at the end of the workshop. We started to see changes in how the children were drawing their whales, using a lot more dark colours, including background information with an ecological awareness of the environment the whales were living in; the children asked more questions about whales (which we did not always know the answer to, but we did some research – "Is a whale-shark a whale or a shark?" was a memorable question). We gave the children a small pot of soil in which we had planted three crocus bulbs to take home with them, so they could feel some connection between the bulbs at home and the bulbs under the ground at school.

Following the second workshop the school contacted us to tell us that they noticed that the children put a circle of chairs around the grass where we had planted the bulbs to stop people from walking across it; their parents used that area as a shortcut to collect their children at the end of the school day, and the children wanted to stop them. When the chairs were moved, they put up signs telling their parents to stop walking on the grass. Even though the bulbs had not grown and were still safe underground, the children had started to develop a protective relationship with the whale.

The third workshop focused on the children's narrative stories about whales. They told us creative and imaginal stories about what the whales would say if they could talk with us about climate change, how the whales would tell us what was happening under the surface of the sea and the ice. They wrote stories back to the whales about how they felt about climate change and what it was doing to the planet. Their drawings following this final workshop showed dramatic changes. None of the children used the template drawing of a whale we had provided at the start of the project. They drew their own versions of whales and frequently included themselves in the drawings, swimming alongside whales in the ocean, sitting on the back of a whale, exchanging stories with whales, and drawing whales outlined by flowers.

The older children demonstrated a more complex metaphorical under-standing, but all the children showed some symbolic understanding of the relationship created between themselves, whales, and climate change. One child wrote:

> The vulnerability of such a cavernous creature can be very unsuspected. When we usually think of unharmful animals the first thing that hatches into our minds are tiny animals. We never identify that one of the largest mammals ever can be so undestructive and harmless. I think it was so cleverly thought of that we planted bulbs to represent the whale. I personally think that the bulbs indicate a metaphor for how fragile the whale is and how fragile the bulbs are, and we are with climate change.

In a sad but illustrative postscript to the project that mirrors many of the attitudes to nature that have led us to this point in history where we face multiple crises in climate change and biodiversity losses, we observed that the school groundskeeper drove a vehicle across the whale while it was in full flower. They could easily have driven around it. The children witnessed someone act destructively and fail to take care of, or protect their whale, by driving through the middle of it. It was heartbreaking to see such a lack of care from adults, in stark contrast to the children's protection and care for the whale even before it grew above ground. We did not have the opportunity to talk about this with the children as the school had broken up for the holidays. Their living sculpture that they described as "harmless" was harmed and to us this mirrored the lack of care of humans against the natural world, even in a protected environment such as a school playing field (or marine park where illegal fishing takes place). We could hardly believe the metaphor that unfolded in front of us.

Youth Theatre Group – Giving Children and Climate Change a Voice

The Free Association Narrative Interview method developed by Hollway and Jefferson (2013) was used to inspire the workshops with youth theatre groups, with the intention that by using open-ended narrative questions I could collect data less conditioned by familiar discourse and capture internal conflicts and underlying attitudes that are not easily accessed consciously. This methodology particularly suits research with defended subjects, and climate change is a subject that people understandably defend against. When used with children, it allows the children to set their own agenda and gives them freedom to follow different narratives. While it could be argued that children could be a less defended group than adult participants, they are also strongly influenced by social, parental, and family attitudes; so, while maybe not individually as defended, I wanted to allow for the fact that they may be mirroring the general "defendedness" that exists in society about the subject of climate change. Romanyshyn (2013) encourages the researcher to "play" with the imaginal landscape of the work, use active imagination, let go of the work (or at least let go of the ego attached to the work), use a reflective process of reverie to establish dialogue between the conscious and unconscious mind of the researcher, and stay uncertain. Free association was encouraged in the workshop narratives which involved eliciting stories structured by "unconscious logic" and feelings rather than "rational intentions" (Hollway & Jefferson, 2013, p. 37). Free Association Narrative Interviewing lends itself to research with children as it can be used playfully when talking about sensitive subjects such as climate change, so is arguably more appropriate for younger research participants and, using creative methods, addresses many of the ethical concerns about researching with younger children (Tidsall et al., 2009; Darbyshire et al., 2005).

Personification was key to the research on thoughts and feelings about climate change with children. Personification is the attribution of a personal nature or human characteristics to something that is non-human or the representation of an abstract quality into human form, as part of being human (Bering, 2006; Kwan & Fiske, 2008). Personification describes an embodiment or symbol, image or representation of an object or thing in usually human form but may also be animal or plant, which can help us move into relationship with the object and develop deeper meaning through bringing things to life (Stewart, 1998). Sometimes argued to be a projective technique, it has roots in narrative therapy, gestalt therapy, psychodrama, and Jungian or transpersonal psychotherapy, imagining something in human form in order to form a relationship with it. Personification is frequently used in literature, poetry, and psychotherapy as it includes attributes of feelings, metaphoric speech, personality, and character "as if" human. In psychotherapy, it draws on the theory behind multiplicity of the person that there are many parts to a person. Problems can be turned into a person, which then allows a dialogue to be developed between person and problem (Rowan, 2010).

Personification is potentially adaptive because of its transformative nature, by changing an event or situation from an external obstacle into an internal meaning. Used here to talk about climate change it created a space in which the children and young people were able to play and show their thoughts and feelings about climate change without leaning on science or factual knowledge. Metaphor and active imagination used as part of the psychosocial research process can help to address what Macaskie and Lees (2011) describe as the "defensive failures" of research by helping to examine the reality of the inner world rather than staying on the surface of the relationship (p. 412).

Active imagination is a technique used to amplify images that emerge in the process of psychotherapy (Jung, 1960; Schaverien, 2005). I am interested in research as a relational process using active imagination and metaphor that can "mobilise the psyche" (Schaverien, 2005, p. 127) and so enable the participants to respond in ways that may not be entirely linear or rational but are imaginal and maybe access unconscious thoughts and feelings.

Working with a youth theatre group (two groups of 20 children), I used the single research question from my qualitative research with individual children (Hickman, 2019):

> *If climate change was an animal or a plant, what would it be, and what would it say?*

Many different images were produced by the children, including very loud lions, dragons, sea monsters, giant snakes, tigers, and wolves. I have conducted over 300 individual interviews with children in many different countries over the last eight years using this single question, and many of the same animals and plants reappear. What was unique about the workshops with the youth theatre group was the appearance of semi-humans, hybrid humans and animals, and creation stories alongside faceless humans and alien plants that no one knew the names of or how to cope with. After the children were invited to personify climate change with this question, they were then invited to enact the voice of climate change as represented by the creature they had imagined it to be. They were encouraged to make noises, interact with each other if they wished to, and speak about climate change. A number of children created stories about horror films they had some vague awareness of; the younger children voiced a distrust of humans and people who deny the problem of climate change; some children turned the stories into humans becoming the predators and destructive impulses affecting animals and plants and children themselves: "You are doing this to the planet, to children and to animals, so we are doing this to you now, as revenge." The alien plants took back control of the earth to protect the animals from humans who did not care about the future of the planet; some children strongly identified with animals facing extinction, saying that they had no future on earth, the same as tigers and polar bears. Narratives of revenge, destruction, and cruelty were prominent, which echoed my experience using this method in individual interviews with children (Hickman, 2019).

> *The father's eyes just grew bigger and bigger as he listened to his son happily paint a metaphorically graphic picture of the interrelationship between climate change, consumption, economic growth, polluting destructive industrial practices, the juxtaposition of consumption and destruction, death, and extinction.*
> *(Hickman, 2020, p. 12)*

At the end of the workshops, the children were supported to disidentify from their personification of climate change as an animal and to say goodbye to the animals they had embodied and listened to. All the children were encouraged to comment on what they had heard and learnt from the animals to enable them to return to their own human identity and think about how they would tell their family about the workshops later as a further distancing or grounding exercise. We did not want the children to return home identified with the animals they had been interacting with. The animals were all thanked and encouraged to also return home to their lands and tell their families about their adventures talking with human children. The children who had imagined hybrid humans/animals/aliens/monsters sent them home to their imaginary planets or lands. They were confident that these creatures knew how to get home. It was important to give care to the children as well as make sure they had the opportunity to talk through any feelings they had following the workshops with us before they returned home themselves. Comments from the children suggested that they felt relieved to be able to talk about climate change, they had fun even though sometimes they were talking about difficult things, they wanted to repeat the exercise the following week, they told us that they were going to get the animal to help them talk with their parents about climate change, and they wanted to repeat the exercise in school with other friends.

Talking to the Trees Project – Psycho-educational Model

An in-school workshop on climate anxiety centred on talking with trees. The aim was to support children (ages 11–15) on an emotional journey discussing climate change through conversations with trees in their school grounds and surrounding green spaces. The children had all volunteered to attend the workshops and in a preparatory discussion they acknowledged that they were feeling anxiety about climate change. They had data-driven and scientific knowledge about climate change but told us that they had very few opportunities where they could discuss how they felt about this; some had friends who could understand their concerns, but some had friends who were dismissive. Some told us they had lost friends who were angry at them for talking about climate change, and others had experienced other children becoming so upset and scared that they had to change the subject and stop talking about it, and they felt guilty towards their friends. Some children told us they could talk with their parents, but others did not.

All the children were members of a school eco-club, planning eco-activities in the school. But to date the focus had been very practical in nature such as growing vegetables and reducing plastic bottles or paper napkins in the school – there were no safe or protected spaces to talk about feelings. Positive emotions of "hope" and "optimism" were promoted as empowering feelings for "doing something" to tackle climate change. But the children told us that they wanted to also talk about the other more difficult feelings they felt including fear, anxiety, sadness, terror, and anger. There were conflicting views on the value of hope and optimism, with some children in the group arguing that they wanted to feel some hope, and others arguing that hope was an obstacle to climate action and another form of climate denial.

All the children self-identified as feeling climate anxiety, but none of them perceived this as a problem in itself; they all saw climate anxiety as a reasonable and understandable emotional response to the pressures of knowing about climate change and its impact on the world. They wanted help to deal with the feelings which were sometimes difficult to cope with, and find more ways to talk with other children about climate change and climate anxiety without scaring them or unleashing fears that they could not cope with themselves. None of the children wanted to get rid of the feelings as they agreed that this would be a further betrayal of the realities of climate change, but they wanted to develop more understanding of how to address the feelings they had that were sometimes hard to articulate and made them feel isolated from their peers. Also they wanted to discover ways to help other children talk about climate change without being overwhelmed.

I used a psycho-educational approach which included the following: The children were asked to hold a conversation with a tree in which they told the tree about their own hopes and fears about climate change and climate action, if they could hear the tree's reply they should include that in the story. They were asked to return to the group with a narrative story that they could share with the group. This allowed the children to begin to develop a shared narrative about their feelings about their individual and collective needs when talking about climate change. Because of the different perspectives in the group, with some leaning towards hope and optimism, and others strongly advocating for more difficult painful truths, it was important to create a shared understanding of all the different feelings and needs in the group and to help them understand that they all had their place. I present this as an emotional biodiversity (Hickman, 2024). Psychologically, in terms of the group, it was important to address differences, defences, and splits in the group as a step towards a coherent and accepting culture in which differences could all be contained and respected, and in which both ecological and emotional biodiversity could be appreciated. Once the children understood emotional diversity, a group culture developed that was more tolerant of difference and they could see that these differences gave the group strength, rather than weakness.

I was then able to further educate the children about mental health and the value of nurturing an emotional range or emotional intelligence. I taught them about climate psychology and the different defences that people hold to protect themselves against climate anxiety so that they could tackle the defences they met in other people's responses and reactions when they tried to talk about their own feelings about climate change. I introduced an adaptation of Fisher's Transition Curve (see Figure 12.2) developed as a visual model that can help children and young people "think the unthinkable and make sense of the range of feelings they are experiencing" (Hickman, 2023, p. 186) which maps the range of climate emotions and shows how to move through them, transforming climate/eco-anxiety into eco-understanding, eco-empathy, and eco-meaning.

Through using this change curve model, the children were able to gain a deeper understanding of their own feelings, explore how these feelings appear in others, and rather than label feelings as good or bad the model helped children to understand and value the mixture of feelings they experienced. The children felt empowered to cope with their own feelings, continue to take action through their school eco group, and also draw on a range of psychological and social skills to help them in the future.

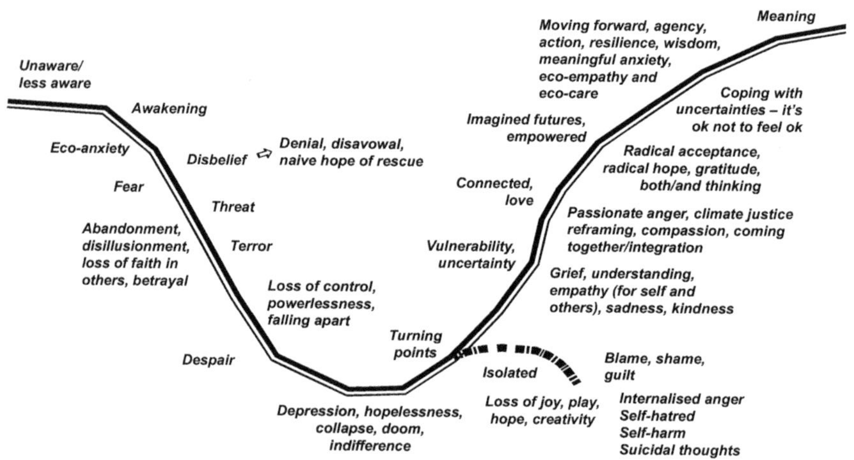

Figure 12.2 Climate Change Emotional Transition Curve

Talking to the Birds Project

Project conducted with Dr. Chris Pawson (UWE, Bristol).

Working with a group of 12 children aged 10–12, we wanted to explore ambivalent and mixed feelings about animals in relation to climate change. People tend to divide animals into groups of those we want to save and protect (whales, bees, dolphins, polar bears), and those we perceive as a nuisance or pest and are disposable (sharks, wasps, bugs, mosquitos). Using Free Association Narrative Interviews (FANI) with individual children and in the youth theatre project, as outlined earlier, some key themes emerged including "empathetic relationship with (non-human) nature and climate change personified as a destructive vengeful force" (Hickman, 2019, p. 55). Working with a puppeteer, the children were taught to make a seagull puppet over a few weeks. Seagulls are a protected species; however, in the local area they are perceived as a pest because they steal food on the streets, divebombing, and attacking people; they make nests on rooftops and make a lot of noise; they have been known to attack small pets; and anyone eating an ice cream on the streets is living a dangerous life. Seagulls have moved inland away from coastal areas in part due to a decline in the fishing industry from overfishing and climate change which has warmed the seas, driving their food source into deeper cooler waters. The gulls have had to find alternative sources of food and the Georgian buildings of Bath in the UK resemble their clifftop homes at the coast. Climate change means that conflicts between people and animals are only likely to increase, with polar bears, elephants, foxes, and walrus all making news headlines as they move into urban areas.

We used the puppets as a vehicle for imaginal storytelling. We educated the children about seagulls, created the puppets, and then, rather than interview the children about climate change and gulls, we interviewed the puppets. The children used their seagulls to talk about how climate change was affecting the gulls, and the gulls talked about how they felt when they were so unwelcome in the cities. We filmed the children and puppets in front of a green screen and then made short films of the narratives. At the end of the project, the children had developed a different understanding of the pressures on gulls facing climate change and saw that human-caused climate change was going to lead to more and more human–wildlife tensions. The children saw it as crucial that we considered the impact of climate change on animals as well as people and felt empathy and concern for the gulls towards whom they had felt ambivalence or antagonism at the start of the project.

Conclusion

This chapter outlined several projects conducted with children and young people in schools/education settings exploring their feelings and thoughts about climate change using art, stories, and theatre. Starting from the position that climate anxiety is an emotionally mentally healthy response to the pressures the planet is facing, we feel this anxiety because we care about the planet, so we should feel proud that we care, and by extension, therefore, feel proud that we feel climate anxiety. To attempt to fix climate anxiety, remove it, manage it, ignore, deny, or avoid it would be fruitless (because climate change will continue to worsen unless radical swift action is taken globally) and denial of their feelings is the opposite of what children need. Children need to be seen, heard, and understood when they are imagining futures that include the realities of climate change. They also need creative ways to explore their feelings and thoughts without being further traumatized. Haraway (2016) argues that we must learn to live, think, see, and feel differently in the face of the climate crisis; learning and teaching differently about climate anxiety could be argued to be central to this. In much of the modern world, children have been separated from the natural world (Griffiths, 2013; Duncan, 2018; Dolan, 2022) but the children quoted at the start of this chapter were clear about their desire to reconnect with nature, with feelings, with relationships, politics, and activism in relation to climate change. They see this as the education they need for the future.

References

Anderson, J., Staunton, T., O'Gorman, J., & Hickman, C. (Eds.). (2024). *Being a therapist in a time of climate breakdown*. Routledge.

Bering, J. M. (2006). The folk psychology of souls. *Behaviour and Brain Science, 29*(5), 453–462. https://doi.org/10.1017/S0140525X06009101.

Darbyshire, P., Schiller, W., & MacDougall, C. (2005). Extending new paradigm childhood research: Meeting the challenges of including younger children. *Early Childhood Development and Care, 175*(6), 467–472. https://doi.org/10.1080/03004430500131247

Davenport, L. (2017). *Emotional resiliency in the era of climate change: A clinicians guide*. Jessica Kingsley Publishers.

Dolan, A. (Ed.). (2022). *Teaching climate change in primary schools. An interdisciplinary approach*. Routledge.

Duncan, R. (2018). *Nature in mind: Systemic thinking and imagination in ecopsychology and mental health*. Routledge.

Griffiths, J. (2013). *Kith: The riddle of the childscape*. Penguin Books.

Haraway, D. J. (2016). *Staying with the Trouble: Making Kin in the Chthulucene*. Duke University Press.

Hickman, C. (2019). Children and climate change: Exploring children's feelings about climate change using free association narrative interview methodology. In P. Hoggett (Ed.), *Climate psychology: On indifference to disaster* (pp. 41–59). Palgrave Macmillan. https://doi.org/10.1007/978-3-030-11741-2_3

Hickman, C. (2020). We need to (find a way to) talk about . . . Eco-anxiety. *Journal of Social Work Practice, 34*(4), 411–424. https://doi.org/10.1080/02650533.2020.1 844166

Hickman, C. (2023). Feeling OK with not feeling OK: Helping children and young people make meaning from their experiences of climate emergency. In L. Aspey, C. Jackson & D. Parker (Eds.), *Holding the hope: Reviving psychological and spiritual agency in the face of climate change*. PCCA Books.

Hickman, C. (2024). Climate aware therapy with children and young people to navigate the climate and ecological crisis. In J. Anderson, T. Staunton, J. O'Gorman & C. Hickman (Eds.), *Being a therapist in a time of climate breakdown*. Routledge. https://doi.org/10.4324/9781003436096-18

Hickman, C., Marks, L., Pihkala, Clayton, S., Lewandowski, E., Mayall, E., Wray, B., Mellor, C., & van Susteren, L. (2021). Climate anxiety in children and young people and their beliefs about government responses to climate change: A global survey. *Lancet Planetary Health, 5*(12), E863–E873. https://doi.org/10.1016/S2542-5196(21)00278-3

Hoggett, P. (Ed.). (2019). *Climate psychology: On indifference to disaster*. Palgrave Macmillan.

Hohti, R., & Karlsson, L. (2014). Lollipop stories: Listening to children's voices in the classroom and narrative ethnographical research. *Childhood, 21*(4), 548–562. https://doi.org/10.1177/0907568213496655

Hollis, J. (1996). *Swamplands of the soul. New life in dismal places*. Inner City Books.

Hollway, W., Hoggett, P., Robertson, C., & Weintrobe, S. (2022). *Climate psychology. A matter of life and death*. Phoenix Publishing House Limited.

Hollway, W., & Jefferson, T. (2013). *Doing qualitative research differently: A psychosocial approach*. Sage.

Jung, C. G. (1960). *On the nature of the psyche (from collected works vol. 8)*. Bollingen Foundation.

Kwan, V. S. Y., & Fiske, S. T. (2008). Missing links in social cognition: The continuum from nonhuman agents to dehumanized humans. *Social Cognition, 26*(2), 125–128. https://doi.org/10.1521/socol.2008.26.2.125

Lawrance, D. E., Thompson, R., Fontana, G., & Jennings, D. N. (2021). *The impact of climate change on mental health and emotional wellbeing: Current evidence and implications for policy and practice*. Grantham Institute Briefing Paper No. 36, 1–36. https://spiral.imperial.ac.uk/bitstream/10044/1/88568/9/3343%20Climate%20change%20and%20mental%20health%20BP36_v6.pdf

Lawton, G. (2019). I have eco-anxiety but that's normal. *New Scientist, 244*(3251), 22. https://doi.org/10.1016/S0262-4079(19)31914-1

Lertzman, R. (2019). *Climate psychology: On indifference to disaster*. Palgrave Macmillan.

Macaskie, J., & Lees, J. (2011). Dreaming the research process: A psychotherapeutic contribution to the culture of healthcare research. *British Journal of Guidance & Counselling, 39*(5), 411–424. https://doi.org/10.1080/03069885.2011.621523

Marks, E., & Hickman, C. (2023). Eco-distress is not a pathology, but it still hurts. *Nature Mental Health, 1*, 379–380. https://doi.org/10.1038/s44220-023-00075-3

Mathers, D. (Ed.). (2021). *Depth psychology and climate change: The green book*. Routledge.

McLeod, A. (2008). *Listening to children: A practitioners guide*. Jessica Kingsley Publishers.

Nelson, C., Bhutta, Z., Harris, N., Danese, A., & Samara, M. (2020). Adversity in childhood is linked to mental and physical health throughout life. *BMJ, 371*, m3048, 1–9. https://doi.org/10.1136/bmj.m3048

Obradovich, N., Migliorini, R., Paulus, M. P., & Rahwan, I. (2018). Empirical evidence of mental health risks posed by climate change. *Psychological and Cognitive Sciences, 115*(43), 10953–10958. https://doi.org.10.1073/pnas.1801528115

Ogunbode, C., Doran, R., Hanss, D., Ojala, M., Salmela-Aro, K., van den Broek, K. L., Bhullar, N., Aquino, S. D., Marot, T., Schermer, J. A., Wlodarczyk, A., Lu, S., Jiang, F., Acauadro Maran, D., Yadav, R., Ardi, R., Chegeni, R., Ghanbarian, E. Somayeh, Z., . . . Karasu, M. (2022). Climate anxiety, wellbeing and pro-environmental action: Correlates of negative emotional responses to climate change in 32 countries. *Journal of Environmental Psychology, 84*, 101887. https://doi.org/10.1016/jenvp.2022.101887

Pihkala, P. (2020). Anxiety and the ecological crisis: An analysis of eco-anxiety and climate anxiety. *Sustainability, 12*(19), 7836. https://doi.org/10.3390/su12197836

Pihkala, P. (2022). Toward a taxonomy of climate emotions. *Frontiers in Climate, 3.* https://doi.og/10.3389/fclim.2021.738154

Randall, R. (2019). *Climate anxiety or climate distress? Coping with the pain of the climate emergency.* https://rorandall.org/2019/10/19/climate-anxiety-or-climate-distress-coping-with-the-pain-of-the-climate-emergency/

RCPsych. (2023, November). *Eco distress for parents and carers.* www.rcpsych.ac.uk/mental-health/parents-and-young-people/information-for-parents-and-carers/eco-distress-for-parents?searchTerms=climate%20anxiety

Romanyshyn, R. D. (2013). Making a place for unconscious factors in research. *International Journal of Multiple Research Approaches, 7*(3), 314–329. https://doi.org/10.5172/mra.2013

Rowan, J. (2010). *Personification: Using the dialogical self in psychotherapy and counselling.* Routledge.

Schaverien, J. (2005). Art, dreams and active imagination: A post-Jungian approach to transference and the image. *Journal of Analytical Psychology, 50*(2), 127–153. https://doi.org/10.111/j.0021-8774.2005.00519.x

Stewart, W. (1998). *Dictionary of images and symbols in counselling.* Jessica Kingsley Publishers.

Tidsall, E., Kay, M., Davis, J. M., & Gallagher, M. (2009). *Researching with children & young people; research design, methods and analysis.* Sage.

UNICEF. (2021). *The climate crisis is a child rights crisis: Introducing the children's climate risk index.* United Nations Children's Fund. www.unicef.org/media/105376/file/UNICEF-climate-crisis-child-rights-crisis.pdf

Verlie, B. (2022). *Learning to live with climate change. From anxiety to transformation.* Routledge.

Weintrobe, S. (2013). *Engaging with climate change: Psychoanalytic and interdisciplinary perspectives.* Routledge.

Weintrobe, S. (2021). *Psychological roots of the climate crisis. Neoliberal exceptionalism and the culture of uncare.* Bloomsbury.

Wray, B. (2022). *Generation dread. Finding purpose in an age of climate crisis.* Penguin Random House.

13 Final Thoughts for a Path Forward

Julie K. Corkett and Astrid Steele

From the outset, we imagined this book as an exploration of children and youths' responses to a critically changing climate, how it is reshaping the planet and impacting our everyday lives. We wondered specifically if students are reacting emotionally to information and experiences of climate change, and how educators are responding, if at all, to their students' expressions of concern. It is clear that many children and youth around the world are very concerned about their present and their futures as framed by climate change. Terms such as climate anxiety and eco-anxiety have entered the language of mental health and well-being, including in educational settings. Feelings of anxiety are not pathological but rational responses to the real and significant threats posed by climate change (see Chapters 3, 4, 8, and 12).

Indeed, the chapters in this book reveal the dual challenges of teaching *about* climate change while simultaneously addressing the emotional impact the content may have on students. Certainly, it is essential that teachers ensure that their students have a strong academic understanding of the processes of climate change and have opportunities to mobilize their knowledge. But in addition, addressing climate anxiety in the classroom requires a holistic and multi-faceted approach that integrates pro-environmental behaviours, helps students internalize climate-friendly norms, and employs engaging teaching strategies (see Chapter 6). Drawing from the chapters, this conclusion synthesizes teaching strategies that can both enhance students' engagement with environmental issues and help mitigate feelings of helplessness and anxiety. As illustrated throughout the book, by fostering a supportive and dynamic learning environment, teachers can empower students to confront their climate anxiety constructively and become active participants in environmental stewardship.

Climate Change Knowledge

To prepare teachers for addressing climate change, a comprehensive curriculum is required that equips students with the knowledge, skills, and values needed to navigate and address the complexities of climate change (see Chapters 2, 6,

DOI: 10.4324/9781003494416-15

and 8). Students' knowledge about climate change needs to include an understanding of the science of climate change, as well as how anthropogenic environmental impacts such as industrialization, deforestation, and pollution have significantly altered the planet's climate. Teaching students about the historical context of climate change may help students understand the direct link between human actions and environmental degradation, thereby fostering a sense of responsibility and agency; without a robust understanding of climate change, students may experience increased uncertainty and confusion.

When teaching about anthropogenic environmental impacts consistency and clarification of all terminology related to climate change and emotions are important, as misunderstandings and misconceptions can exacerbate anxiety, they can lead to unnecessary fear and worry. Teachers should encourage students to ask questions and seek clarification, and they should provide opportunities for interactive discussions, fact-checking activities, and critical analyses of media sources. By providing accurate information and debunking myths, teachers can help students develop a more grounded and realistic understanding of climate change. Furthermore, by regularly revisiting and defining terms such as climate anxiety, eco-anxiety, sustainability, and resilience, students can feel more comfortable and confident in their knowledge rather than be distressed by confusion and uncertainty.

In addition to fostering clarity in terminology and understanding, integrating the history of the environmental movement can provide students with a broader context for addressing current environmental issues and students' climate anxiety. The historical timeline of the environmental movement should include key moments such as the publication of Rachel Carson's *Silent Spring* in 1962, the first Earth Day in 1970, Greta Thunberg's speech at the UN's 2019 Climate Summit, governmental environmental policies such as the Clean Air Act in the United States and international agreements such as the Kyoto Protocol and the 2015 Paris Agreement, and the regular reports of the IPCC (Intergovernmental Panel On Climate Change). An understanding of key policies and agreements is pivotal for students' understanding of the complexities of climate governance, the challenges in policy implementation, and the importance of international cooperation. Furthermore, learning about these milestones may help students appreciate the evolution of environmental awareness and the efforts made to mitigate climate change and its attendant issues.

Governmental policies and international agreements that address climate change are often tied to economics (see Chapters 6, 9, and 10). Therefore, it is important for students to explore alternative economic systems that are vital to developing a holistic understanding of sustainability. Traditional market-driven economies often prioritize profit over environmental health, leading to unsustainable practices. In contrast, alternative economic models, such as the circular economy, which promotes sustainability through recycling, reusing, and reducing waste, may reduce climate anxiety as it inspires innovative thinking and encourages students to envision a more sustainable future that values relationships and community over mere economic gain. The exploration of the economic connection to climate change must include the intersectionality of ecological, social, and economic dimensions of sustainability. For example, students need to understand that climate change can exasperate social inequalities, as vulnerable communities are often the most affected by environmental degradation. Highlighting these interconnections may help students understand the multifaceted nature of sustainability and the need for integrated solutions.

Critical Thinking and Problem Solving

When students feel capable of contributing locally to climate solutions, they are less likely to feel overwhelmed by the magnitude of global climate change (Chapters 3, 6, and 9). The sense of agency and empowerment derived from knowing that they can make a difference will help mitigate students' feelings of helplessness and despair, as is apparent throughout the book. Presenting a balanced perspective of climate change by combining discussions of climate threats with stories of hope, resilience, and activism can mitigate feelings of futility. This approach not only strengthens students' self-efficacy by showcasing real-life examples of overcoming challenges but also helps them manage their emotions more effectively. Balancing negative and positive narratives ensures that students remain motivated and optimistic about their ability to contribute to climate solutions.

Students should be made aware that learning about, and experiencing climate change may not be the main, or only factor contributing to their anxieties. Students need opportunities to explore how their concerns are influenced by various social and public media, and how those media may affect perceptions of future risks (see Chapters 3, 5, 6, 9, and 11). Examining the impact of social media can help students to understand public attitudes, behaviours, and anxiety towards climate change. For example, media literacy education can teach students to critically assess the portrayal of climate issues in various media and effectively learn to source balanced perspectives. In this age of information overload, AI-generated misrepresentations, and false narratives, it is important that students learn to distinguish between credible sources, misinformation, and disinformation. To achieve that ability, students require foundational knowledge in critical thinking and research skills to evaluate scientific data, understand the impacts of climate change, and explore potential solutions (see Chapters 2, 7, 10, and 11). Empowering students with the ability to critically analyse information will enable them to make informed decisions and engage in effective climate action.

Teachers can enhance students' critical thinking and problem-solving skills by providing students with opportunities to analyse environmental issues and develop potential mitigations to climate change. Learning opportunities such as assignments that require students to research, propose, and implement sustainability initiatives not only teach relevant environmental science but also develop skills for addressing the complex challenges associated with climate change. When students research, propose, and implement sustainability initiatives, they experience firsthand the process of turning ideas into actionable solutions. Hands-on learning helps students see the tangible impact of their efforts and goes a long way in reinforcing their belief that they can make a difference. As students encounter and overcome obstacles during such projects, they practice the creative process of design, of articulating and communicating ideas, and implementing their ideas effectively, their confidence in their problem-solving abilities grow. Seeing the positive effects of their initiatives within their communities or school environments can provide a sense of accomplishment and reinforce their commitment to environmental stewardship. Ultimately, as argued in Chapters 2, 10, and 11, by integrating critical thinking and problem-solving activities, teachers enhance not only students' knowledge about climate change but also their confidence and emotional resilience.

Interactive and Engaging Strategies

Woven throughout the book is a call for teachers to use interactive and engaging teaching strategies that foster an appreciation for environmental sustainability and promoting a sense of agency and connectedness. Student-directed inquiry allows students to take ownership of their learning and explore topics of personal interest; when students are actively involved in their learning, they are more likely to feel empowered and capable of making a difference. By exploring topics that resonate with them personally, students can discover ways to connect their actions to broader environmental goals, fostering a proactive rather than a reactive stance towards climate change.

Evident throughout the chapters is the importance of designing lessons that incorporate sustainable practices and encourage students to adopt pro-environment behaviours in their daily lives. For example, science lessons can include ways to reduce carbon footprints, recycling initiatives, and sustainable energy generation design. By connecting pro-environmental behaviour to developmental goals such as leadership development, teachers can provide their students with the opportunity to lead initiatives, organize events, and influence their peers, all of which enhance leadership skills and agency. By making these behaviours part of the regular classroom routine, students can see the tangible impact of their actions, thereby invalidating feelings of helplessness and fostering a sense of agency.

As argued in Chapter 10, incorporating environmental literature that emphasizes connection, interdependence, and environmental ethics encourages students to reflect on their own relationship with nature and others, which can deepen students' emotional engagement and understanding. Literature that highlights the interconnectedness of all living things can help students see the broader impact of their actions and the importance of collective efforts in addressing climate change. By using interactive and reflective teaching strategies, teachers can create a learning environment that not only educates students about environmental sustainability but also addresses their climate anxiety. Students who feel connected to their learning and to the natural world are more likely to develop a resilient and hopeful outlook, and, in turn, will be better equipped to cope with the emotional challenges connected to climate change. This can reduce feelings of isolation and anxiety by reinforcing their understanding that they are part of a larger community working towards common goals.

Addressing Morals and Ethics

Students often perceive behaviours related to the environment as either morally right or wrong; however, morals and ethics are rarely so clear-cut. To help develop an understanding of the complexity of environmental ethics, students need to explore the relationship between climate change and social and environmental injustices. Chapter 4 explores how students' understanding of moral injury, moral distress, and moral residue can grow their awareness of the ethical issues inherent in human-induced climate change. Students, for example, can examine how vulnerable populations, often those least responsible for environmental degradation, are disproportionately affected by the impacts of a climate that is increasingly unpredictable and chaotic. By examining social injustices, students learn to navigate their own ethical and moral responses to climate change, and perhaps develop strategies for their emotional resilience (see Chapters 4 and 9). Recognizing the vulnerabilities of certain groups with respect to a changing climate demonstrates empathy and social awareness, which are key components of emotional intelligence. By addressing ethical issues associated with climate change in a supportive and empathetic manner, educators help vulnerable students feel understood and reassured, which can alleviate anxiety and promote a sense of agency. This approach may ensure that all students, regardless of their background, feel empowered to take part in environmental activism.

It is important to help students identify their personal motives for environmentally associated behaviours – whether positive or negative. Students need to recognize that their personal norms and values do not develop in isolation from society; social dynamics, such as a desire for acceptance and conformity, can drive students to adopt behaviours of their peers – which may or may not be environmentally friendly. Therefore, teachers are encouraged to include discussions on the role of advertising, consumer culture, and peer pressure in promoting unsustainable environmental practices. By crucially examining these influences, students can become more conscious of their own behaviours and choices – behaviours and choices based on their growing ethical perspectives, which in turn may mitigate their climate anxieties.

As discussed in Chapters 2, 4, and 5, teachers are encouraged to help students internalize climate-friendly personal norms by connecting pro-environmental behaviours to the students' existing values. When doing so it is essential that teachers consider the students' cultural, social, and individual backgrounds, and then frame environmental responsibility in a way that resonates with students on a personal level, challenging their beliefs and lived experiences. Diverse and inclusive materials can provide different perspectives and enrich the learning experience. For instance, by assisting students to develop a sensitivity to local environmental issues teachers can engender an understanding of the importance of far-reaching and global ecological interactions. Educators can design projects that investigate local environmental challenges, such as water pollution, deforestation, or urban heat islands thus connecting global climate issues to local contexts; students can see the immediate relevance of their actions and feel more motivated to contribute to solutions.

Climate-Related Emotions

Young people, regardless of their age, often react with strong emotions, with passion, to the information and experiences that they encounter. Regardless of their students' ages, it is important for teachers to understand that students may not have sufficient understanding of their strong emotions, of why they feel worried or anxious about issues such as climate change. It is important to teach students about the psychological mechanisms behind worry and anxiety, such as distinguishing between rational concern and pathological anxiety. Rational concern involves a healthy awareness of environmental issues and a commitment to addressing them, whereas pathological anxiety can lead to debilitating fear and inaction. Educating students about psychological mechanisms may help them demystify their emotional experiences and thereby promote healthier coping strategies.

Drawing from multiple chapters, we find a general agreement that teachers need to foster students' empathy and ethical compassion towards all living things as a foundational step in addressing climate anxiety. By cultivating empathy, students can develop a deeper connection to the natural world and a sense of responsibility towards its preservation. However, as students develop greater "nature" empathy, there is an increased risk of students experiencing an emotional response to climate change. Therefore, recognizing the range of possible emotional responses to climate change is essential for fostering emotional resilience. Students may experience a spectrum of emotions, such as anxiety, anger, hope, fear, optimism, and pessimism, and these emotions can significantly influence a student's willingness to engage in climate action. For instance, while anxiety may be paralysing, anger can be a powerful motivator for change. Helping students to navigate their emotions constructively may enhance their ability to engage in sustained climate action.

Students often feel powerless in the face of climate change due to their limited capacity to influence change, which can exasperate climate anxiety (see Chapters 3, 6, 8, 9, and 12); therefore, it is important that students develop coping strategies for their emotions. In Chapter 7, Wiseman et al. suggest that coping strategies be categorized into problem-focused, emotion-focused, and meaning-focused approaches. Problem-focused coping involves taking actionable steps such as participating in community cleanup projects. Emotion-focused coping emphasizes managing emotional responses to climate change through practices like mindfulness and other stress reduction techniques. Meaning-focused coping refers to finding personal significance in environmental action, which can provide a sense of purpose and motivation. All three strategies reflect teaching and learning strategies that foster the creation of meaning, hope, engagement, and self-efficacy, all of which may go a long way in addressing students' climate anxiety. Incorporating strategies that help students practice different coping mechanisms will enhance their emotional intelligence and resilience; using various coping strategies, students can better manage their emotions, reduce anxiety, and improve their overall well-being. Also, including content on youth activism and ways to influence policy, such as voting, advocacy, and participating in environmental organizations, can further empower students. By providing students with an understanding of these platforms and how they can be used to voice concerns and engage in political processes, students may find themselves engaged in activities that mitigate their feelings of helplessness.

Activities

The second section of the book provides detailed practical examples of activities that can help students develop an understanding of climate change and simultaneously address students' climate anxiety. We offer a summation and some additional activities that can be developed and modified to meet the needs of individual classrooms, students, and communities.

(1) Provide opportunities for students to share their knowledge and feelings about climate change and environmental sustainability with their community, Indigenous Elders, political leaders, and parents.

(2) Have students attempt to solve real-world environmental challenges through research projects that will deepen students' understanding of climate change and promote active learning.

(3) Engage in discussions and debates about current environmental policies, such as

　　a. The impacts of climate anxiety and its role in motivating actions versus its potential to cause distress. This can help students see multiple perspectives and develop their own informed opinions.

     b.   The moral responsibilities of individuals versus collectives in addressing climate change.

(4) Have students engage in local conservation projects and community-based environmental initiatives.

(5) Collaborate with local governments, organizations, and community members to enable students to contribute to environmental protection efforts in their community.

(6) Create discussions or reflective exercises where students can explore and discuss their feelings about climate change, and in which students are encouraged to use precise terminology to articulate their emotions.

(7) Conduct role-playing or simulations wherein:

     a.   Students can practice political advocacy and environmental activism, helping them feel more empowered to take actions.

     b.   Students simulate decision-making processes of political leaders and climate activists.

(8) Have students reflect on their own carbon footprints and the emotions associated with their choices (e.g. feeling guilty for taking a car vs. feeling proud for using public transportation).

(9) Have students create a workshop to teach others about climate activism, including practical steps that can be taken to make a difference.

(10) Hold a Youth Climathon (www.unescap.org/events/2024/youth-climathon-innovative-solutions-acceleration-climate-action-11th-apfsd-online).

(11) Make a multimedia presentation examining the lived experiences of people from different parts of the world who are impacted by climate change.

(12) Analyse case studies that present moral dilemmas related to climate change, encouraging students to evaluate different perspectives and develop their own reasoned conclusions.

(13) Use art, poetry, plays, storytelling, or music to enable students to express their feelings about climate change.

(14) Establish climate clubs that focus on both activism and emotional support. Climate clubs can provide a space that is free from the constraints of academic performance in which students can share their experiences and strategies for coping with climate change.

(15) Conduct a climate impact analysis project in which students assess the environmental impact of common behaviours in their community.

(16) Invite guest speakers who are involved in sustainability efforts to share their experiences.

(17) Use a values mapping exercise in which students identify and reflect on their core values and how these can guide pro-environment behaviour.

(18) Create a photovoice activity to encourage individual expression and self-determination of pro-environmental behaviour.

Conclusion

Throughout this book, it has been argued that it is the responsibility of educators, policy makers, and community members to create learning environments that support the holistic development of students, empowering them to become informed, resilient, and ethically responsible global citizens. The incorporation of climate change knowledge, mental health and well-being, and ethical/ moral considerations into one's teaching is not just an educational imperative, but a societal one. By providing students with a comprehensive understanding of climate change, fostering their mental resilience, and creating opportunities to develop strong ethical values, we can prepare students to face the challenges of the future. The implementation of appropriate teaching strategies enables teachers to promote a supportive and dynamic learning environment that addresses climate anxiety and fosters resilience, critical thinking, and a commitment to environmental stewardship. An integrated approach to climate change education ensures that students are not only informed about the scientific aspects of climate change but also equipped to manage their emotional responses, engage in ethical decision making, and take meaningful action. Through a comprehensive and empathetic approach, students can be empowered to navigate the complexities of climate change and contribute positively to a sustainable future.

Index

For Product Safety Concerns and Information please contact our EU
representative GPSR@taylorandfrancis.com
Taylor & Francis Verlag GmbH, Kaufingerstraße 24, 80331 München, Germany